EXPL⬤RE!

A
of SCIENCE

Explorations in Science

PROGRAM AUTHORS

STEPHEN CAMPBELL
DIANA KAYE GOOLEY
LALIE HARCOURT
DOUG HERRIDGE
GILLIAN KYDD
SHERRY MAITSON
NANCY MOORE
BEVERLEY WILLIAMS
RICKI WORTZMAN

EXPLORE!
A BOOK of SCIENCE

JAY INGRAM

DOUG HERRIDGE
NANCY MOORE

Addison-Wesley Publishers Limited

Don Mills, Ontario • Reading, Massachusetts • Menlo Park, California
New York • Wokingham, England • Amsterdam • Bonn
Sydney • Singapore • Tokyo • Madrid • San Juan

Editorial Development *Susan Petersiel Berg* • *Susan Hughes* • *Keltie Thomas*
Photo Research *Nyla Ahmad* • *Louise MacKenzie*
Design and Art Direction *Wycliffe Smith*
Electronic Production *Wycliffe Smith Design Inc.*
Cover Photograph *Michael Dick, Animals, Animals*

Acknowledgments for the selections appear on page 232.
See page 236 for a complete list of photographers. See page 237 for a complete list of illustrators.

Scientific Reviewers
Dr. Peggy Foegeding, Associate Professor, Department of Food Science, North Carolina State University
Dr. Karen Goodrowe, Metro Toronto Zoo
Dr. Allan Griffin, Department of Physics, University of Toronto
Dr. Wayne Hawthorn, Department of Biology, University of Waterloo
Harold Hosein, Meteorologist, CITY TV, Toronto
Dr. A.G. Lewis, Acting Head, Department of Oceanography, University of British Columbia
Dr. Margaret Silliker, Department of Biology, De Paul University
Sandra Van Nooten, St. Paul's School 'Woodleigh,' Baxter, Victoria, Australia
Dr. Mason R. Yearian, Department of Physics, Stanford University

We are grateful to the following consultants:
Charles Barman, Ed.D., Associate Professor of Science Education, Indiana University
David Brummet, Ph.D., Math and Science Education Consultant, Palo Alto, California
Michael A. DiSpezio, M.A., Science Education Consultant, Cataumet, Massachusetts
Vallie Guthrie, Ph.D., Director, Greensboro Area Math and Science Education Center, North Carolina Agricultural
and Technical State University
Michael B. Leyden, Ed.D., Professor of Education, Eastern Illinois University
Sheryl Mercier, M.A., Elementary Science Specialist, Fresno Unified School District, California
Karen Ostlund, Ph.D., Associate Professor of Education, Southwest Texas State University

Canadian Cataloguing in Publication Data

Ingram, Jay
Explore! : A book of science, 5

Includes index.
ISBN 0-201-56068-2

1. Science — Juvenile literature. I. Herridge , Doug, 1954 —
II . Moore , Nancy , 1952 —
III . Title .

Q163.I54 1993   j500   C92-095559-2

ISBN 0-201-56068-2

A B C D E F — BP — 98 97 96 95 94 93

CONTENTS

Dear Reader,

If you want to see what's extraordinary about the world around you, take a look at the world as scientists see it.

You can see a solar-powered weapon that may have destroyed an invading army in ancient Greece. You can check out tiny flowers that follow the sun in the Arctic. You can see how animals have become world-class travelers. You can even get a taste of summer on Saturn!

You might think these things come from someone's vivid imagination, but they don't. They are all stories about our world, uncovered by scientists. Go ahead: turn the page and start seeing the world through the eyes of scientists.

Jay Ingram

Go WITH the Flow

WHAT'S AHEAD

JUST Coiling Around

BY JAY INGRAM

The next time you're lucky enough to have pancakes with maple syrup, watch what happens when the syrup hits the pancakes.

If you hold the bottle high enough, you'll see the syrup coil up like a snake before it flows into a smooth puddle. Why does syrup coil? Milk doesn't coil … orange juice doesn't. But honey does and even ketchup does, sometimes. So what's going on?

Syrup is thick — if it starts to run off the sides of your pancakes, you've got time to scoop it up and put it back on top. Scientists say syrup is a viscous (VISS-kuhs) liquid, meaning it's thick and doesn't flow very easily. If you could actually take a photo of the tiny syrup molecules (MALL-uh-kyools) — a very magnified photo — you would probably see that the molecules have a hard time moving past each other when syrup flows.

(A molecule is just a group of atoms that are joined together.) They bump into each other, or get tangled up. In fact, ice-cold syrup hardly flows at all. But when syrup is heated, it flows more easily. That's because the syrup molecules bounce around and vibrate more when they're heated. So the syrup molecule traffic jam starts to thin out.

What happens when you take syrup out of the cupboard and pour it over your pancakes? The syrup doesn't spread over the pancakes right away, because it has a hard time flowing. So it starts to pile up.

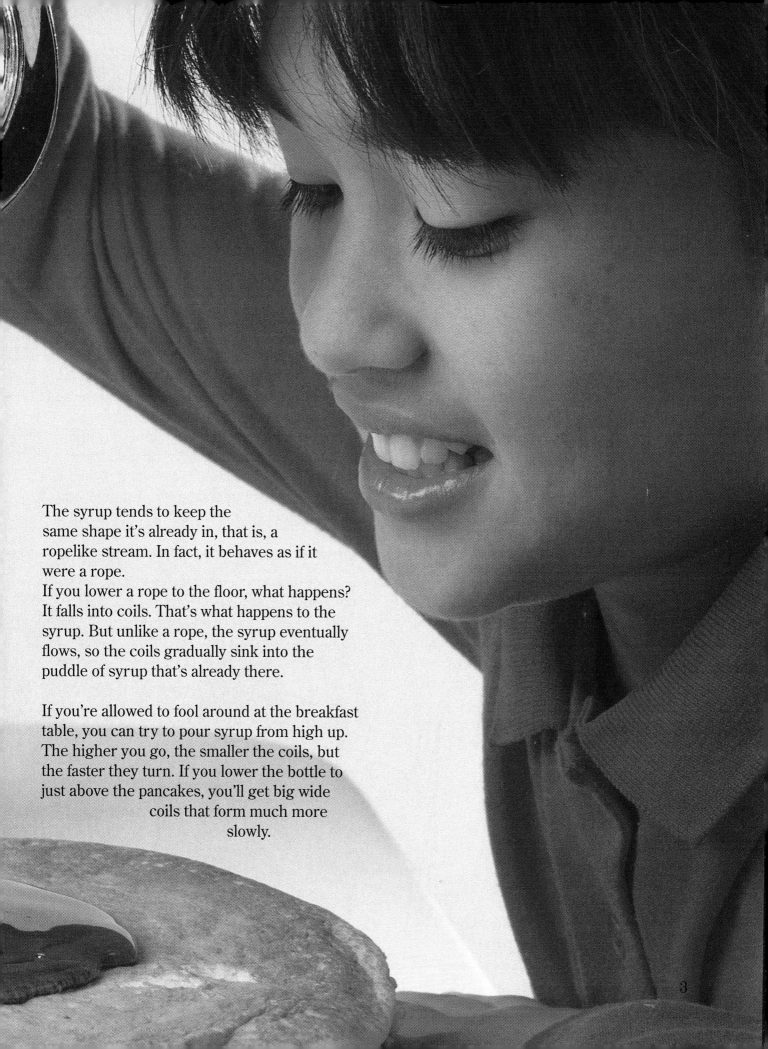

The syrup tends to keep the
same shape it's already in, that is, a
ropelike stream. In fact, it behaves as if it
were a rope.
If you lower a rope to the floor, what happens?
It falls into coils. That's what happens to the
syrup. But unlike a rope, the syrup eventually
flows, so the coils gradually sink into the
puddle of syrup that's already there.

If you're allowed to fool around at the breakfast
table, you can try to pour syrup from high up.
The higher you go, the smaller the coils, but
the faster they turn. If you lower the bottle to
just above the pancakes, you'll get big wide
coils that form much more
slowly.

3

Ketchup is even stranger than maple syrup. If you pour ketchup onto your french fries from a height of two to five centimetres, it'll form nice, even coils. But if you raise the bottle and keep pouring, the coils disappear! The ketchup hits the french fries and just flows over them. Why does this happen?

You know how you have to smack a ketchup bottle to get the ketchup out? The more you hit it, the easier it flows. Somehow giving ketchup molecules a shock makes them slip by each other more easily. When you pour ketchup from two to five centimetres high, the impact of landing isn't much. So the ketchup stays stiff, and you get beautiful coils. But if you lift the bottle up and then pour, the ketchup lands with a jolt big enough to shock it into flowing.

SECRET SAUCE

Why does ketchup need a good shake to get it out of the bottle? Because it's thixotropic. Thixo-what? A thixotropic (thik-suh-TROH-pik) liquid is a liquid that's set like a jelly until you disturb it. When you shake ketchup, you disturb it and break its setting. After that, it flows out freely.

A few years ago nondrip paints appeared. They look like jelly in a can. When you dip a paintbrush into

4

Suddenly — no coils!

The next time you're fixing yourself a bite, why don't you try comparing coils from barbecue sauce, mayonnaise, and a thick chocolate milkshake? Think of the delicious snack you could have. Think of the mess you'd make in the kitchen … and you could tell your parents it's all in the name of science!

the jellylike stuff, the paintbrush comes out covered with sticky lumps. But these lumps paint on just like slippery liquids.

Cans of nondrip paint have DO NOT STIR clearly marked on them, because if you stir the paint, you'll break the jellylike setting. Then it will drip and run over you or anything else that's in its way, just like ordinary paint. So as long as you remember not to stir it, you don't have to worry about splattering drops of paint everywhere. Now all you have to worry about is tripping over the paint can!

Try **THIS**

Try gently lowering a heavy rope to the floor. Watch it coil. Find other things that behave this way.

Water doesn't coil like maple syrup and ketchup. But it does flow in a ropelike stream. The next time you're brushing your teeth, watch how the water flows out of your tap.

GET THE LOWDOWN
on
ooze

BY JIM KILLGORE

Amazing fluids that defy even Newton!

At 11:00 this morning, a box appeared on my desk. How long I'd waited for this moment. Laughing, I tore open the box. Inside were three cans. Immediately, I ripped the lid off one and held it open over my ancient manual typewriter. Slowly, a green glob oozed over my typewriter, eventually swallowing the keys.

Slimy fingers spilled onto my desk, working their way toward my stapler.

Could this actually be part of my job? Under a tight deadline, my assignment was to explain the science behind the gunk that was then oozing over my pencil. It seems that some

6

fluids, such as toy-store ooze and putty, differ from everyday fluids in weird ways. For example, ooze will bounce, pour in a ropelike flow (that you can actually cut with scissors), pull itself from one container into another, or snap into pieces if you yank it.

All of this struck me as pretty odd for a fluid. Water doesn't bounce like a ball. I've never seen syrup pull itself, or siphon, onto a plate of pancakes. I needed answers — fast!

My first step was to call physics professor Jearl Walker. Walker will do just about anything to illustrate a point, including sticking spoons to his face and breaking boards with his bare hands. Walker uses ooze, slime, gunk — call it what you will — as props in his lectures.

Walker told me that ooze is a good example of what's called a non-Newtonian fluid. "Ah, that's interesting," I said. "How about telling me what a Newtonian fluid is — so I at least know what kind of fluid ooze is not."

Walker sighed and decided he'd better take things from the top. "Let's define a fluid," he said. "A fluid is anything that flows. It could be a gas, a liquid, an avalanche, or a bunch of logs rolling together down a hill."

All fluids, he continued, have a property known as viscosity (viss-KAWSS-it-ee). It tells you how thick a liquid is, or how much it resists flowing (the thicker a liquid is, the more it resists flowing). Ketchup's viscosity is what makes it ooze rather than spray across the counter when you shake it over your hamburger.

7

Newtonian fluids follow a law of viscosity first explained by physicist Isaac Newton. The law says that the viscosity of a fluid can be changed only by changing the fluid's temperature. "A common example," said Walker, "is honey. You warm up honey and it flows more easily." (That means the honey becomes less viscous.)

Well, a non-Newtonian fluid breaks Newton's law. Its viscosity can also be changed by applying force or stress. In other words, if you slap, punch, or throw ooze against a wall, its viscosity increases — tremendously. It won't splatter or flow. It will act like a solid — a piece of tough elastic. Yank the ooze and it will snap like a rubber band. Throw it on the floor and it will bounce.

Walker's explanation left my mind swimming with questions. Unfortunately, he had a lecture to teach. But he gave me the number of a brilliant chemist — a regular ooze expert — and wished me luck.

As it turned out, this chemist had done lab studies on the many wacky fluids in toy stores. I found out that ooze is made with guar (gwar) gum, a natural substance found in the seeds of an Asian pea plant. Its molecules look like globs of sticky, cooked spaghetti. The tangled glob that the long and stringy molecules form is the key to its weird behavior.

Left to flow, ooze molecules will slide over each other, gently untangling — like cooked spaghetti being poured from a pot. The fluid flows with a fairly low viscosity.

WHAT'S IN THE TUBE?

Is toothpaste a liquid or a solid? If you squeeze the tube, the paste flows like a liquid. If you leave it alone, it sits there in a lump like a solid. What do you think it is?

Toothpaste is a plastic liquid. Plastic liquids flow only when they're pushed. Take the top off the tube, and you can wait all day for the stuff to run out. But if you squeeze the tube, then out it squirts.

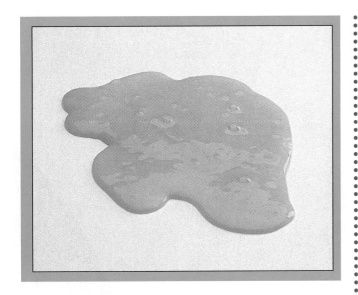

But when ooze is yanked or pounded, these long, stringy molecules form instant knots. Some of the knots snap like rubber bands. Others draw back against the force, giving the ooze a rubbery feel. Viscosity instantly increases which means that the ooze flows less. And under extreme force, ooze will act like a solid — and actually break. However, once the force is removed, the knots gently loosen and the viscosity returns to normal.

After filling me in on ooze, the chemist chatted with me about a few other non-Newtonian fluids — such as cornstarch in water and quicksand. But I was anxious to get off the phone and get my fingers back into the real stuff.

Try making your own wacky fluid. Pour 175 mL of water into a bowl. Slowly stir in 500 mL of cornstarch. Make sure your mixture is thick. If the mixture isn't thick enough, you might end up with a big, splattering mess. Check the thickness by pushing your hand down into the stuff. Add more cornstarch if needed. Grab onto the fluid and pull it out of the bowl. Try rolling it into a ball, bouncing it, stretching it, pouring it, and snapping it.

Glass is not a solid. It's a very thick liquid that flows so slowly that it keeps its shape for many years. That's why old windows are thicker at the bottom than the top. Over the years, the glass has flowed down like thick molasses.

The BLUE Planet

BY KELTIE THOMAS

Someday, you and your friends may travel through space. As you speed toward the stars, one of the most amazing sights you'll see is our world. Oceans surround all the land and flow over three-quarters of the Earth. And there's even more water hidden beneath the land, in layers of loose sand and gravel.

Although Earth has so much liquid water that it looks blue from space, water is a rare liquid. You won't find it anywhere else in our solar system. You'll find frozen water on Mars and in some other places. You'll even discover some water that's gas, or vapor. But you won't see a drop of liquid water beyond Earth.

As you shuttle among the other planets in our solar system, you won't see any sign of life, either. Scientists think that's because none of the other planets have any liquid water. They believe that Earth's first forms of life developed in the ocean millions of years ago. When creatures evolved to live on land, their bodies carried part of the ocean ashore. Did you know that almost three-quarters of your body is made up of water?

10

11

Earth wasn't always such a watery world. Scientists think our planet formed about 5000 million years ago. But it wasn't until almost 1000 million years later that water appeared. Even then you wouldn't have seen swirling blue oceans, because this water was vapor.

A water molecule is made up of two hydrogen atoms and one oxygen atom.

Water vapor formed when the invisible gases hydrogen and oxygen combined. When two hydrogen atoms join to an oxygen atom, a molecule of water forms. (A molecule is just a group of two or more atoms that are joined together.) H_2O is the formula for water. It tells you that two hydrogen atoms and one oxygen atom make up a water molecule.

As more and more water vapor formed, it collected in the atmosphere — the thin layer of gases surrounding Earth. Sometimes, the vapor cooled and formed raindrops. But Earth was so hot then that these raindrops never reached the ground. As they fell through the atmosphere, they were heated and became vapor again. Gradually, our planet cooled. Then raindrops could fall to Earth and fill hollow valleys and cracks to form oceans, lakes, rivers, and streams.

Of course, water has never stopped changing form. On Earth, you're surrounded by invisible water vapor that hangs in the air. It forms when sunbeams heat a body of water. When this vapor cools, liquid water falls out of the sky as rain. Or liquid water freezes, disguising itself as hail, snow, or ice. Nothing else transforms itself as easily as water.

In the winter, the three forms of water — mist, liquid, and ice — exist all at once at Niagara Falls.

If the water in Earth's oceans were spread out evenly over the Earth, we would be covered by a layer of water more than three kilometres deep.

CHECK IT OUT!

Open an atlas and take a look at a map of the world. Why do you think many towns and cities have sprung up near lakes and rivers? What parts of the world have little or no fresh water?

In some places where fresh water is scarce, desalination (dee-SAL-ih-NAY-shun) plants take salt out of seawater to make it drinkable. Find out how countries such as Kuwait desalinate seawater to make fresh water.

You'd need to take lots of water for your journey in space. Although you could live for several weeks without any food, you could only survive for a few days if you ran out of water. That's because water does some amazing things that all people, plants, and animals depend on for survival. Water combines with, or dissolves, the nutrients that are in food and carries them to where they are needed in living things. After that, living things can turn nutrients into energy that can be used for growth and repair. Water also helps get rid of wastes. When wastes are dissolved in water, they can be carried away. All of these processes can only happen in water.

In the spring, ice begins to melt and break up on the St. Lawrence River.

Mist curls around the trees in a forest.

So don't forget to look at planet Water if you find yourself roaring through the galaxy one day. Wait a second — was that planet Earth, or Water? What would *you* call a planet that's practically covered in water and populated by creatures that can't survive without it?

ROUND AND ROUND

Water is heavy. When Space Station "Freedom" orbits the Earth, it will be able to hold only three months' worth. And it'll cost too much to launch a shuttle to carry up more. So scientists are developing ways to recycle all the water that's on board — again and again. That way the crew members will never need more than they start out with!

When astronauts sweat or breathe, they release water. This water can be collected and condensed in loops on the walls or ceiling of the station. (Urine, shower water, and washing water will be recycled separately and used only for showers and washing.) Then the collected water can be filtered a few times and boiled to remove tiny particles and microbes. Iodine is added to get rid of any surviving microorganisms. Voilà! Pure drinking water.

Endless Journeys

Did you know that the water that flows from your tap may have been drunk by a dinosaur millions of years ago? That's because water is recycled in nature. Follow the journey it takes.

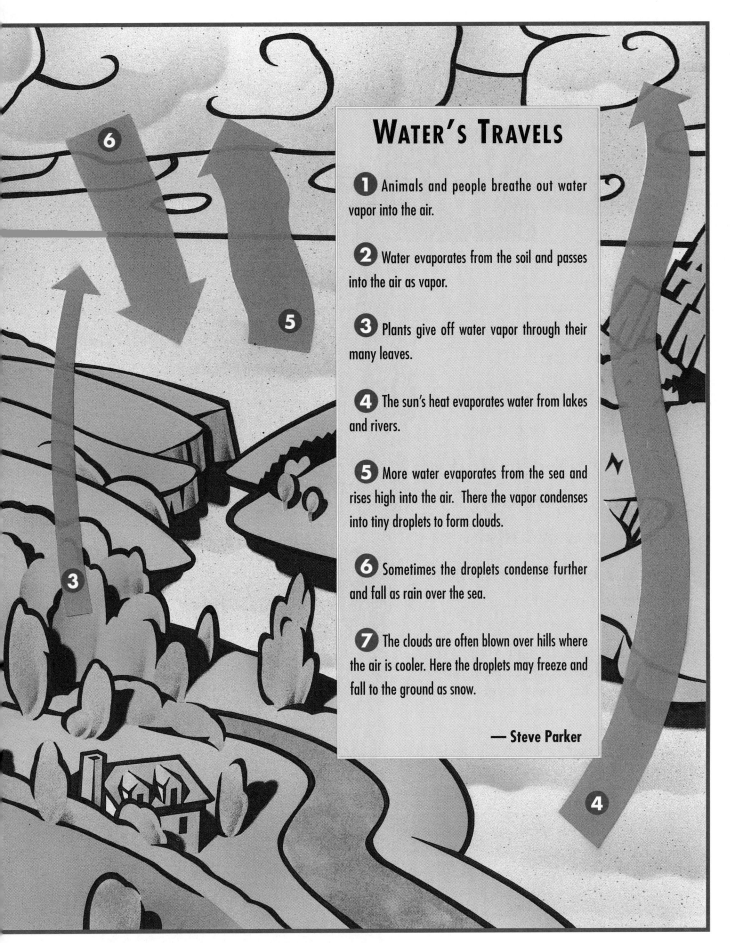

WATER'S TRAVELS

1 Animals and people breathe out water vapor into the air.

2 Water evaporates from the soil and passes into the air as vapor.

3 Plants give off water vapor through their many leaves.

4 The sun's heat evaporates water from lakes and rivers.

5 More water evaporates from the sea and rises high into the air. There the vapor condenses into tiny droplets to form clouds.

6 Sometimes the droplets condense further and fall as rain over the sea.

7 The clouds are often blown over hills where the air is cooler. Here the droplets may freeze and fall to the ground as snow.

— **Steve Parker**

The Goo in You

BY VICKI COBB

To stare
a watery
creature in the eye,
you don't need to find a
shark, a whale,
or even a goldfish.
All you need is a mirror.

Now you might think you're pretty dry, but that's because your skin is the driest part of you. Inside, your body is wet and gooey.

Water is in every part of you. It's in liquids like tears, sweat, blood, saliva, and urine. Water is also in your muscles, and bones, and even your brain. In fact, there is more water in you than any other single substance. If you weigh 27 kg, about 19 kg of your body is water. That's more than half of you — no wonder your insides are gooey!

Most of the water you have taken into your body has become a part of you. It is inside your cells, the smallest living parts of you. Inside cells, water is not runny or even watery. It is more like jelly.

Your body needs to take in water to make fluids. The fluids your body makes that are most like water can pass out of cells. Tears, sweat, and urine are not pure water because they contain salt. Urine also contains other wastes that would poison you if they were not carried out of your body.

Your saliva is stickier and slimier than plain water. It needs to be because of the job it does. When you chew your food, you mix saliva with it. Saliva helps the food move more smoothly down your esophagus (ee-SAW-fih-guhss) — the long tube leading to your stomach. It also holds the chewed food together, making sure it all reaches your stomach and doesn't collect in your mouth or esophagus.

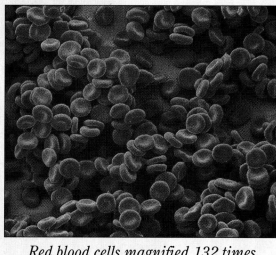

Red blood cells magnified 132 times

19

Saliva also gets food ready for your stomach. It starts breaking down some foods so that they can be taken into cells after digestion. For example, starch molecules in crackers, potatoes, and other foods are too big to enter cells. Saliva changes them to sugar molecules that can be taken into cells.

The inside of your nose and the tubes that lead to your lungs, called tracheae (TRAY-kee-ee), are lined with another kind of goo, mucus (MYOO-kuhss). Thick, slimy mucus protects you from dust and microbes in the air. It acts like flypaper — dust

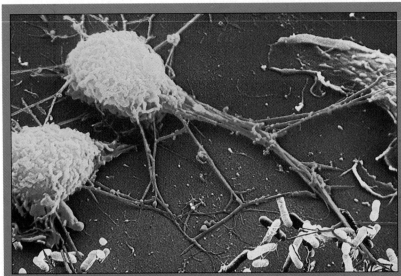

White blood cells surround microbes, magnified 2150 times. The microbes have been colored yellow.

and microbes get stuck in it. Sometimes, the microbes get past the mucus into your cells and you catch a cold. Then your cells react by making lots of mucus. The extra mucus tries to flush the microbes out of your head. That's why you get a runny nose when you have a cold.

If you have ever had a cut that became infected, you've seen another kind of goo, pus. Pus is a yellowish-white, thick, creamy material. It shows that your body has been fighting the microbes that have infected you.

A FRESH START

Kevin Bell puts water into a large jar with a gravel filter. He collected this water from a storm drain that empties into the Stillwater National Wildlife Refuge in Nevada. After he learned that many birds were dying there, Kevin set up a science fair project to find out why.

Kevin studied fish and plants

When you cut yourself, blood comes out. Microbes, or tiny organisms, can enter the cut. When this happens more blood rushes to the wound to fight the microbes. This is why the area around the cut swells and becomes red and sore.

Your blood contains white blood cells which surround and destroy microbes. Many white blood cells die from destroying microbes. Pus forms when the dead white blood cells mix with some of the liquid part of your blood.

If you clean a cut right away, it won't become infected and your white blood cells won't have to fight microbes. White blood cells are one way your body protects you against microbes. It makes extra white blood cells when there is a danger of infection.

Some people may think the goos in your body are disgusting. Now you can tell them all the important jobs that gooey stuff does for your body.

in the refuge. He found that minerals in the water from the irrigation of fields were poisoning the wildlife in the area. Kevin discovered that adding three parts of fresh water to one part of drain water kept the wildlife he was studying alive.

Fresh water is now flowing into the refuge. For his work, Kevin won a President's Environmental Youth Award. But Kevin says his biggest reward is knowing he's helping save some animals' lives.

Try
THIS

Does saliva really change starch into sugar? See for yourself. Chew a plain soda cracker, one without salt. Keep it in your mouth without swallowing for five minutes. Then taste the chewed soda cracker with the tip of your tongue. It should taste sweet.

What do jellyfish and cucumbers have in common? They're both 95 percent water.

FLASHBACK

Ellen Richards was a trend-setter. She was the first North American woman to earn a chemistry degree, the first woman admitted to the Massachusetts Institute of Technology (MIT), and the first president of the American Home Economics Association.

Richards was born in 1842 in Massachusetts. When she received her degree in chemistry in 1873, Richards was interested in the environment and nutrition. She tried to combine her interests by working in the area of public health. She believed that drinking water was often spoiled by sewage waste and other harmful substances. So she looked for ways to test for pollution in drinking water.

Ever wonder where the term "home economics" comes from? It came out of several meetings that Richards led to build a bridge between science and

Trendsetter

the home. She believed that science and the problems of everyday life were closely linked and wrote many books about the connections she saw.

Richards also helped develop a way to improve the smell of waste water. Her method used filters made of straw. When waste water passed through the filters, bacteria in the water would collect on the straw. These bacteria attracted and removed the chemicals in the water that made it smell.

The methods Richards found to purify drinking water and treat sewage have developed into the ones we use today. Because she wanted to train others in the same field, she helped establish a department of sanitary chemistry at MIT. Many of her students went on to help make water and sewage treatment everyday procedures.

Sinking fast

This actor is about to drop in for a "quick" chat, but it doesn't look like he'll be leaving anytime soon. Do you think the quicksand in this old movie is actually like the real stuff?

BY JULIE WILLIAMS

I f you said no, you're right. Just take a look at the photograph on the opposite page. But would you know what to do if you suddenly found yourself in the "quick" of it?

If Indiana Jones found himself kneedeep in quicksand, he'd just grab his trusty bullwhip and lasso the nearest tree branch to pull himself to dry land. You could do just as well armed with a little information.

You won't find the word "quicksand" in a reference book about rocks and minerals, because quicksand isn't a type of sand. It's just ordinary sand that's in a waterlogged condition. In the word "quicksand," "quick" means mobile, or moving.

When water seeps into sand from the bottom, as it can when sand sits on a base of clay that doesn't let water drain through, the grains of sand float and behave more like a liquid than a solid. A heavy object like a rock can sink easily in quicksand, almost as fast as it would in water. Because quicksand is thicker, or denser, than water, you feel like you're being sucked down into it when you step in it.

Obviously, the smart thing to do is to avoid stepping in quicksand in the first place. However, you can't tell just from looking at sand whether it's in this "quick" condition or not. Sometimes, quicksand doesn't look wet because it has a thin layer of dry sand on top of it.

Quicksand usually forms near rivers or beaches. Poke sand in these areas with a stick before you walk on it if you want to check for quicksand. If it's quicksand, you'll see ripples form around the moving stick just as if you were moving a stick through water. Dry sand will not make these waves.

HOW QUICKSAND FORMS

Before

This loosely packed sand is waterlogged.

The "Quick" Condition

When the sand is shaken by an earthquake or by someone walking on it, the space between the sand grains gets larger as water escapes upward. The grains quickly sink and pack together.

After

No longer in the "quick" condition, the sand is tightly packed.

But if you do find yourself sinking fast, the first thing to do is the most important — relax. A little known fact about quicksand is that it's heavier than the human body, so you can't actually sink into it. If you relax and lie down on your back, you will float just like you do on water. Remember, quicksand is just sand float-

Quicksand forms around the Little Colorado River in Arizona.

ing in water. As you lie on your back, wiggle gently from side to side to "swim" to firm ground. If you get tired, simply relax and float for a while. If you panic and thrash around, you could work yourself deeper into the mucky sand to a point where you could actually drown.

Try
THIS

Is it difficult to create quicksand? Give it a try. After all, you only need two ingredients: sand and water.

Keep track of how much of each ingredient you add so you can come up with a quicksand recipe. How does changing the quantities change your results? Does it work better if you stir? How can you tell if what you've made is really quicksand?

Trinidad has a lake filled with asphalt that you can walk right across if you keep moving.

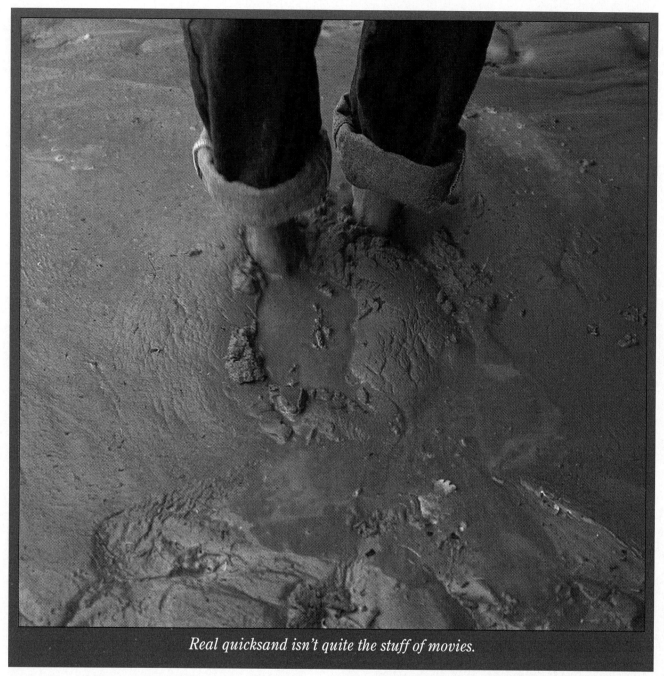
Real quicksand isn't quite the stuff of movies.

If you were with a friend who had managed to stay clear of the stuff, your friend could help you get out. Your friend could throw you a rope and pull you out. Or if you were close together, your friend could lean over to you with a stick and pull you out. In the worst case, you could send your friend for help and relax and float until the help arrived.

Some movies make a big deal out of quicksand. But now you know it's just a mixture of water and sand that acts more like water than sand. When you go walking through a known quicksand area take a stick to test the sand in front of you as you walk. You should do it even if you were just there the day before, because quicksand can form, dry up, and form again.

So if the sand beneath your feet starts to creep around your ankles, keep your cool and relax. Now you know how to get yourself out of it, and you can show those people in the movies how it's really done.

And Now, the Weather

Tracking t

If you accidentally tuned in to an Arctic winter weather forecast, you might think another ice age was coming: "... and this week, we expect daily temperatures to hover around a high of -40°C, with a slight chance of precipitation."

Sun

The dryness and severe cold make growing conditions in the Arctic bad enough, but the weak sunlight that falls there makes them even worse. However, plants have some inventive ways of adapting to the unlikeliest of climates. In the Arctic, you'll find some flowers that are shaped like television satellite dishes.

Take a good look at the next satellite dish you see. Notice how it's curved like a shallow soup bowl? The signal it receives comes from a satellite that's orbiting 35 000 km above the Earth. That's a long way to transmit a TV program! The signal satellites send from so far away is fairly weak and must cover a lot of ground.

So a satellite dish is shaped to get the most out of the signal. It curves inward. Every part of the dish the signal hits, even the edges, bounces the signal to the antenna that's in the middle of the dish. This way the antenna absorbs both the signal that falls directly on it and the signal that all the other parts of the dish receive. That's what makes the signal strong enough to form a picture on your TV.

29

This idea didn't originate with TV engineers. Some Arctic flowers were shaped like this long before the satellite dish was ever invented! They trap sunlight the same way a dish collects signals from a satellite. For most places on Earth, the sun's rays are much more powerful than a satellite's weak signal. But high in the Arctic, the sun is always fairly low in the sky. So the same amount of sunlight that covers a small area at the equator spreads out and covers a much larger area in the Arctic. This means that the sunlight that falls on Arctic flowers is weaker than the sunlight that falls on flowers closer to the equator.

best way to capture as much of the sun's "signal" as possible. Instead of having an antenna sticking up in its middle, the flower has machinery for making pollen and seeds, its stamens and pistil. The warmer these are, the faster they work, which is all-important when time is short.

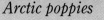
Arctic poppies

Since Arctic summers are very short, flowers need to bloom, be pollinated, and produce seeds in just a few weeks. That's why many of them are shaped like satellite dishes. It's the

Arctic flowers have other tricks up their sleeve too — lots of them turn to follow the sun as it moves across the sky. This ensures that they collect as much heat as possible by pointing directly at the sun for most of the day. Some flowers have very fuzzy stamens. This helps them trap warm air by making it harder for

CHECK IT OUT!

What does the Arctic have in common with the Sahara Desert? Both places are extremely dry. Find out what kinds of plants live in the Sahara Desert. How have they adapted to the hot, dry habitat?

On July 21, 1983, the lowest temperature ever recorded on Earth — a chilly -89.2°C — was recorded at Vostok, Antarctica.

cold air from outside to blow through and cool the flower. By doing all these things, these flowers are able to stay much warmer than the surrounding air — as much as 10°C warmer! This could be the difference between producing seeds in time or not.

The flowers aren't the only Arctic inhabitants that benefit from their clever design. Walk through an Arctic meadow and peek into these flowers, especially in the early morning, and you're likely to find mosquitoes or flies basking in the warmth. These insects need heat to survive. If their flight muscles aren't warm enough, they can't fly. And just like the stamens in the flower, the hairier the insect, the warmer it stays.

Satellite dishes

Flowers are also good places for insects to hide from hungry birds patrolling the meadows for little insect-sized morsels. But if you were an insect, you could never be completely safe … Deep down inside the petals of some of these flowers, where it's very warm, very hungry spiders are hiding, just waiting for lunch to be delivered!

The Arctic may be cold and dry, but it still buzzes with creatures who have found ways to live there.

KEEPING IN TOUCH

Ever seen the wind ripple through a tree? If you think the tree didn't notice, think again. Janet Braam and Ronald Davis have discovered that more than three-quarters of all plants are sensitive to touch. Touched by the wind or a person (it's probably all the same to a plant), within 10 minutes the genes of a sensitive plant react. The biologists think the genes "tell" the cells to grow a sturdier stem. In other words, frequent touching makes the plant grow out instead of up. Braam's not sure why this happens, but she thinks it has to do with survival: a shorter, stockier plant has a better chance of staying in one piece on a windy day than a tall, spindly plant!

Rain

What do tree frogs have in common
with earthworms? You can count on hordes
of them appearing from nowhere when it rains.
But many creatures disappear as soon as the
first raindrops fall. Most flying
insects scramble like mad for
shelter because they don't fly
well in rain. It's even
hard for pilots to fly
airplanes
through
rain.

Raindrops may seem tiny to you, but to an insect they're huge. This stinkbug, or shield bug, is weighed down by a raindrop that's bigger than its head! Drops can cling to an insect's wings. Sometimes they add so much extra weight that the wings become too heavy for the insect to move.

After the rain, this wasp removes water from its nest, one drop at a time. The nest must be kept dry for the wasp eggs to develop.

A flower makes a great umbrella! This antlion clings to the flower's stem and waits to dry. As soon as the sky turns dark and the air feels damp, most flying insects scramble to hide under leaves, flowers, branches, or shrubs. Those that can't find protection from the rain in time hang onto anything nearby.

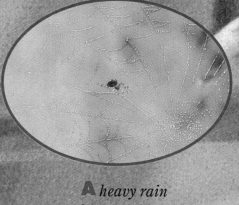

Here's trouble. This dragonfly hasn't found shelter in time, and it's missing a wing. Hope it can hang on till the rain stops. Most flying insects need to keep their flight muscles warm to be able to fly. When it rains, the air cools off. If their flight muscles become too cold to fly, the insects are stranded. So they need to find shelter fast!

A heavy rain can destroy spiderwebs. Even before the web is dry, this spider scurries to repair it. There's no time to lose. Now that the insects are out and about again, a tasty meal might fly by for the snagging any second.

Add **Water** and *Stir*

BY DAVID LAMBERT AND RALPH HARDY

When you wake up to sun streaming through your curtains, you can't wait to go outside. But if raindrops are beating on your window, you might not be in such a hurry.

Weather affects our lives in many ways. It can make or break a vacation. It can help make crops thrive but just as often it can spoil them. Fierce hurricane winds can pluck trees up by their roots. Rough storms can sink ships at sea. Fog can ground planes.

The weather that you find in different parts of the world depends on many things. How hot is the air there? How much water is in the air? How is the wind moving the air? Is the wind lifting the air?

Many years ago, it was discovered that these questions are related to air pressure — the atmosphere's weight at a particular place.

The lower the air pressure is, the more likely you'll find rain and strong winds.

Our planet brews weather much like you might make a cake. First, you put your ingredients in a bowl. Then you stir them up. Next, you bake your mixture in an oven. How your cake turns out depends on four things: your ingredients; how you mixed them; the temperature of the oven; and how long you baked your cake.

The main ingredients of weather are air and water. The three forms of water — ice, raindrops, and vapor — exist in the air. Nature's mixing bowl is the Earth's ball-shaped outer surface. Of course, cake batter would fall out of a mixing bowl shaped "inside out" like this. But air doesn't fall off our planet because gravity attracts it to the Earth's surface. Gravity works like a giant magnet that pulls every object on Earth to the ground.

In India, the beginning of the rainy season is a time to celebrate. Monsoon rains bring relief from the scorching heat of the dry season.

The sun's heat and the spinning of the Earth are like mixing spoons. They stir the air and water about. The sun also acts as nature's oven. It's preset to bake the air each day. The baking time of the air and water varies as the seasons change. In the summer the sun bakes air and water longer than it does in the winter. The heat from the sun also varies from place to place. The closer a place is to the equator, the more direct heat it receives from the sun.

This dry lake at Maryborough, Australia, has been sun-baked to a crisp!

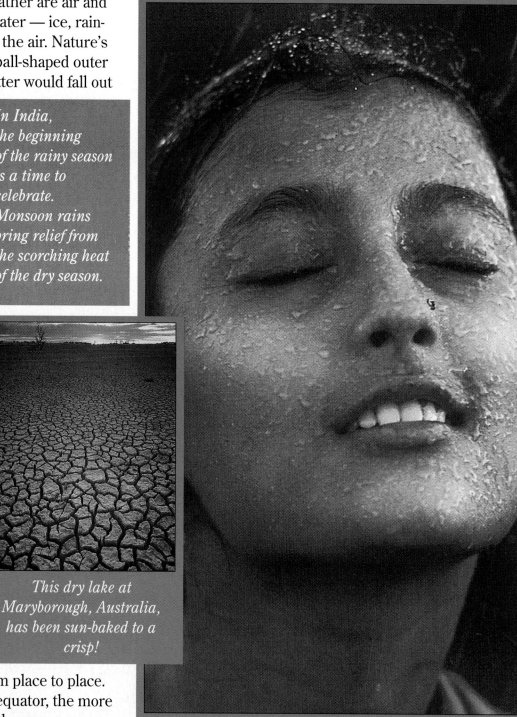

CHECK IT OUT!

If the wild winds of tornadoes, cyclones, or twisters — call them what you will — touch down in your town, look out! They can lift cars and overturn airplanes.

Over 700 tornadoes hit North America every year.

Find out where and when they usually strike and what causes them.

You can bet this cyclist was suprised when a blizzard whirled into Spearfish, South Dakota.

Fog rolls into Petty Harbour, Newfoundland.

Air warmed by the sun combines with the spinning of the Earth to make winds. These winds spread heat and moisture more evenly around the world. Where the warm air cools, you can get clouds, rain, snow, hail, fog, or frost.

Winds are very important because the sun heats the equator much more than the poles. Without winds to help spread heat and moisture, much of the Earth would be impossible to live on.

All kinds of weather happen at once in different parts of the world. Where you live, it might be dull, cool, and drizzly. But somewhere else, the sun is shining while a warm breeze blows. In another place, a blizzards whirls, burying a city in snow. Elsewhere, the sun shines from a cloudless sky over dry, cracked earth. In another corner, heavy rains flood the countryside.

Our weather brews in the lower atmosphere (the atmosphere is an envelope of air that surrounds the Earth). Clouds and dust in the atmosphere absorb or bounce back lots of the energy that the sun rays beam to Earth. So less than half of the rays' energy gets through to warm the Earth's surface and the lower atmosphere.

Somewhere else, fog rolls in from the sea.

Weather is constantly changing. It's what happens in the atmosphere from day to day and hour to hour.

Hurricane winds blow as fast as 119 km/h. But tornadoes have the fastest speeds on record. Some of them have been clocked at speeds above 400 km/h!

A SNOWY DAY IN JULY

Suppose you were lying on a beach in July and snowflakes started to fall. Snow in July? Without even thinking about it, you have come to expect certain kinds of weather at certain times of the year. And snowy days in July just don't fit in.

Your idea of what's normal depends on the climate in your area. For example, if you live in Hawaii, you probably think of January as beach weather. In most parts of Canada and the northern United States you'd have to break through the ice to go for a swim in January. Both are normal.

Each place has its own climate controls, like a computer program, that keep the climate pretty much the same year after year. The most important climate control is how far north or south you live. The closer you are to the equator, the more of the sun's warmth you get. That's why you can take a January swim in Hawaii, but not in Hudson Bay.

Do you live near the ocean or a large lake? Your climate may be cloudier and rainier than that of places farther inland. Your summers will probably be cooler and your winters warmer, too. If you're landlocked, smack in the middle of a continent, the temperature where you live may go up and down more in a single day and from one day to the next than if you live near a large lake. Your summers are likely to be hot and your winters cold. Each area's climate controls shape its weather.

Weather can change overnight, but climate stays the same for many hundreds of years because the same controls are "programming" it. But sometimes the controls change, and when that happens, the climate changes too. The result? Palm trees in the Arctic and, yes, snowy days in July!

— Valerie Wyatt

PREDICTING THE *Unpredictable*

BY KELTIE THOMAS

You won't hear Isabel Ruddick grumble about how unpredictable weather is. That's exactly what fascinates her about it!

Ruddick is a meteorologist. It's her job to forecast the weather. She's part of a team that makes the weather forecasts that are sent to radio and television stations and newspapers.

When Ruddick first arrives at the office, she goes to the briefing room to be briefed on the latest conditions of the atmosphere. Here, she is studying satellite photos of the atmosphere on her computer screen.

Ruddick loves the challenge of trying to predict what will happen in the future. She works with computer models of the atmosphere that predict what the atmosphere will be like in the next 12 hours. Meteorologists call these models "progs" — prog (prawg) is just a short form of the word "prognosis" which means prediction.

Keltie: What do you need to know to forecast the weather?

Isabel: First, you need to know what the weather is right now in the area you're forecasting for. Then you need to know what weather's happening to the west. For example, if it's snowing in Chicago, you know snow's probably going to hit Windsor because weather systems usually move from west to east. Next, you need to know the temperature of the air, the dew point temperature — that's the temperature at which raindrops will form or fall. You also need to know the air pressure, how much cloud is around, and how high the clouds are.

Keltie: How do you collect all of this information?

Isabel: Many ways. Satellite pictures show us where clouds are and how they're moving. We use radar to find out where precipitation is. Radar sends out beams that travel through the atmosphere. If a beam hits a water droplet that's large enough to be rain or snow, it will echo back to us. Then we know that it's raining or snowing there.

Keltie: Can radar tell the difference between rain and snow?

Isabel: No, it can't. We rely on reports to tell us whether it's raining or snowing.

Keltie: Where do these reports come from?

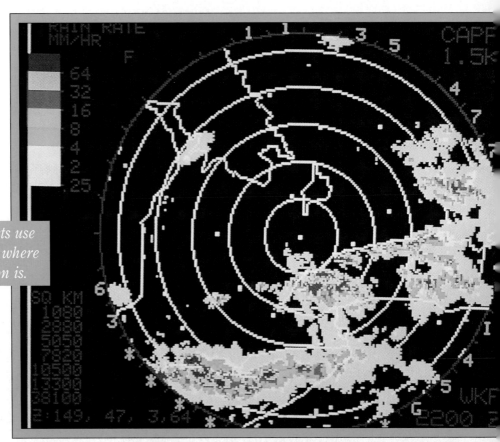

Meteorologists use radar to spot where precipitation is.

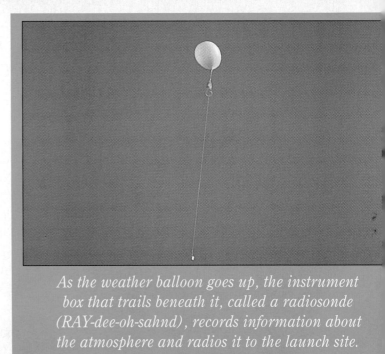

As the weather balloon goes up, the instrument box that trails beneath it, called a radiosonde (RAY-dee-oh-sahnd), records information about the atmosphere and radios it to the launch site.

Isabel: They come from weather stations and autostations. An autostation is a weather station that's able to run on its own — nobody has to be there. It has the same instruments

40

that a person would use to measure the weather such as a thermometer and a barometer. Every hour, it automatically takes these readings.

Keltie: We hear a lot about weather balloons. How do you use weather balloons to collect information?

Isabel: Twice a day, at 7:00 A.M. and 7:00 P.M., weather balloons are sent up into the atmosphere at sites across North America. Every 300 m or so, they measure the air temperature, the dew point temperature, the air pressure, and the speed and direction of the wind. Then they radio this information to weather stations.

Keltie: Once you've collected all of this information, how do you use it to make a forecast?

Isabel: The information collected by the weather balloons is fed into supercomputers at the Canadian Meteorological Centre in Quebec and the National Meteorological Center in Washington. It runs through a program and the computers come up with a prog. The computers give us models of what the atmosphere will be like in a particular area 12 hours from now. We receive printouts of these models. We interpret them, along with the other information that's been collected, to come up with a forecast that's usually — but not always — accurate.

Keltie: Why do you sometimes make mistakes?

Isabel: The computer programs we use are based on observations of past weather. The programs use a simple model of how the atmosphere changes over time. The computer makes some assumptions so that it doesn't

Try THIS

Watch a weather report on television or check your local newspaper for the forecast at the beginning of a week. Write down the forecast for the week. Keep a record of the weather that actually happens each day of the week.

Compare your record with the forecast. How accurate were the predictions? Do this for another week and compare the accuracy of the forecasts for each week.

41

take a whole day to generate the progs. A few things are ignored. That's why forecasts are not always accurate. But I can see them getting better when computers are able to run faster.

Keltie: How accurate are your forecasts?

Isabel: Forecasts for the next 24 hours are usually 80 percent accurate. But forecasts for two to three days ahead are only about 60 percent accurate, and forecasts for five days are about 50 percent — they're not very accurate at all. These long-range forecasts are based only on computer models.

Keltie: What makes weather hard to predict?

Predicting the weather is a team effort.

LOW-TECH FORECASTING

Here are a few old sayings that predict weather. If they always held true, we wouldn't use high-tech forecasting equipment today.

Rain before seven,
Clear by eleven.

Rain may be caused by a rain belt that takes about four hours to pass by. So this saying probably holds true quite often.

Sea gull, sea gull,
Sit on the sand.
It's sign of a rain
When you are at hand.

Air often becomes less thick, or dense, before it rains, so it's less able to support a bird. Since it's harder for them to fly, sea gulls are more likely to stay on land.

Isabel: It's unpredictable! The atmosphere is always changing. Sometimes the computer progs don't really pick up what's happening. Or a small change in the atmosphere may not be detected by the weather balloons, satellites, or radar. If this small thing grows into something big five days from now, then our forecast is going to be inaccurate by that much.

Keltie: Have you ever been really surprised by the weather after you've made a forecast?

Isabel: Oh, yes! Something surprises me every single time. I might have 10 reasons why a cloud should develop and only two for why it shouldn't. So I predict that it will, and then it doesn't. But that's what makes it so interesting!

When ants travel in a straight line, expect rain; When they scatter, expect fair weather.

It's highly unlikely that ants behave like this before it rains. But you can check out the ants for yourself. See if there are any changes in their behavior that you can link to the weather.

— Franklyn M. Branley

Try THIS

Next time a spring storm comes calling, gear up in your boots and raincoat and go on a wildlife hunt. You will find all sorts of small creatures hiding under leaves, bark, and branches. Bring along a magnifying lens in case you decide to study some up close. Don't forget to put each creature back where you found it after you've had a good look.

CHECK IT OUT!

Ask your grandparents what weather sayings they have heard or used to forecast the weather. How reliable do they think each saying is?

World-famous meteorologist Edward Lorenz recently said that long-range weather forecasting cannot be done because the Earth's atmosphere is very sensitive to small changes. A butterfly flapping its wings in Peking might change the weather in New York two weeks later!

BEAT
the ELEMENTS!

Get set to race your friends
across North America!
Try to beat the elements you might
meet along the way —
rain, wind, snow, and other
forces of nature. As you travel,
you'll learn something about
how to react to the powers of nature.

READY? ● *Let's go.*

HOW MANY CAN PLAY?
● Two to six people can play.

THINGS YOU WILL NEED
● One die and a marker for each player.
You can use different colored beans or beads for markers.

HOW TO PLAY
● Roll the die and move ahead the number of spaces shown.
If your marker lands on a space marked with a flag, move ahead
to the first space that has no pictures or words on it.

TO WIN
● The first player to get back to the weather station by an exact
roll of the die wins.

Text within illustration:

19. Sailing on the lake is great now. Move ahead 4 spaces.

20.

21. Fog rolls in. Wait for it lift. Lose 1 turn.

22. Fog clears and the sun comes out. Move ahead 1 space.

23.

24. A tree struck by lightning falls across your path. Lose 1 turn.

25. The power is out. Go back 2 spaces.

26. You're stranded in an area that's flooding. Lose 1 turn.

27. You're rescued by a boat. Move ahead 2 spaces.

28. Evacuate before hurricane strikes. Go back 5 spaces.

WEATHER ON

BY ALAN DYER

Have you ever wondered
why summers are warmer than winters?
Do you think other planets have seasons, too?
Do these other planets even have weather?

Some people think our summers are warmer than our winters because Earth is closer to the sun during the summer. Sounds like a good theory, but you'll discover it can't be right.

It's true that over the year Earth takes to circle around the sun, its distance from the sun does change. (Actually, the path our planet follows is not a perfect circle. It's shaped more like a squashed circle, or an oval.) But Earth is 5 million kilometres closer to the sun in January than July! So this theory can't be right.

> *Earth is closer to the sun in January than in July.*

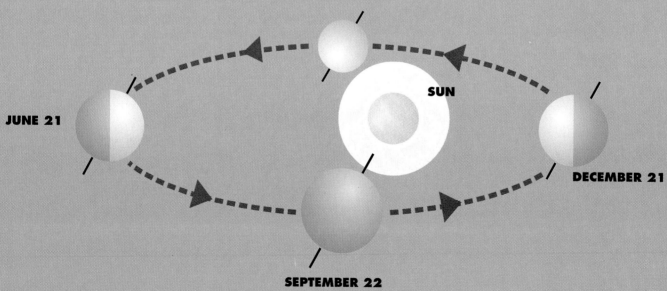

MARCH 21

SUN

JUNE 21

DECEMBER 21

SEPTEMBER 22

The real reason for the seasons is that Earth is tilted. The line, or axis, Earth spins around doesn't point straight up and down; it tilts a bit.

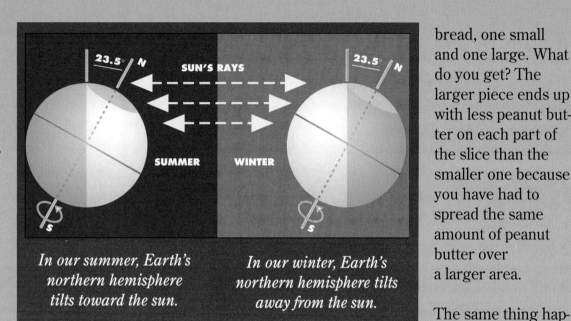

In our summer, Earth's northern hemisphere tilts toward the sun.

In our winter, Earth's northern hemisphere tilts away from the sun.

During our summer, the northern hemisphere of the Earth tilts toward the sun and the sun appears high in the sky. Its rays beam down onto the ground from almost straight overhead. This makes it easy for the sun to heat up this area, and we get the warm days of summer.

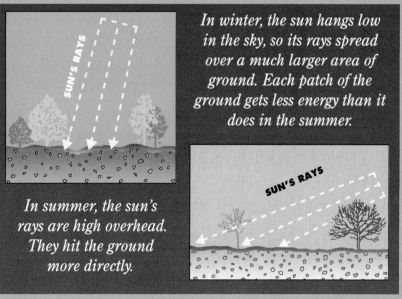

In winter, the sun hangs low in the sky, so its rays spread over a much larger area of ground. Each patch of the ground gets less energy than it does in the summer.

In summer, the sun's rays are high overhead. They hit the ground more directly.

To understand what happens in the winter, think about trying to spread the same amount of peanut butter on two different pieces of bread, one small and one large. What do you get? The larger piece ends up with less peanut butter on each part of the slice than the smaller one because you have had to spread the same amount of peanut butter over a larger area.

The same thing happens to the sun's rays during our winter. The northern hemisphere tilts away from the sun. So the sun always appears low in the sky. The sun's rays spread out over a much larger area of the ground than they do in the summer. Just like the amount of peanut butter, there is less heat on each part of the ground. So we get the chilly days of winter.

VENUS

On Venus, there are no seasons. You won't even have to worry about getting cold at night because its scorching temperature of 460°C (that's twice as hot as your oven when it's turned up high) never changes very much.

Venus is almost the same size as Earth, but it's 41 million kilometres closer to the sun. If you think being nearer to the sun makes Venus hotter than Earth, you're right. But even experts were surprised to find that it's hotter than Mercury, the closest planet to the sun. Venus is so hot because its atmosphere (the layer of gases that covers the planet) is made up of almost nothing but carbon dioxide. This is the gas that could cause global warming on Earth. Venus's atmosphere is much thicker and cloudier than Earth's. The clouds on Venus are so thick that you won't ever see the sun from the planet's surface! The thick atmosphere traps the sun's heat under the clouds and heats up the planet, just like a blanket traps the heat from your body and and keeps you warm at night.

Venus is the hottest planet in the solar system — far too hot for any life to survive.

CHECK IT OUT!

On Neptune, the temperature never climbs above -200°C. Astronomers thought Neptune was too cold to have weather such as storms. But in 1989, photos from the *Voyager 2* space probe showed that it has a dark, stormy spot like Jupiter's Great Red Spot. Go to the library and try to find some photos of Neptune's Great Dark Spot.

Pluto is an icy world that takes 248 Earth-years to travel around the sun. Although its summers last for 40 Earth-years, Pluto never gets warmer than -230°C!

MARS

Get ready to touch down on a more "seasonal" planet. Since Mars tilts like Earth, it has seasons. But Mars is 78 million kilometres farther away from the sun than Earth, and it takes twice as long to go around the sun as Earth does. This means that each season on Mars lasts twice as long as it does on our planet.

Because Mars is farther from the sun, you'll need to bundle up. Even on a sunny summer day, the temperature on Mars is often only -50°C. That's as cold as a winter day in the Arctic. And during a Martian winter, the temperature can drop to a bone-chilling -125°C!

Sometimes, winds sweep over Mars with hurricane force at 200 km/h. Dust storms whirl over the planet as these strong winds pick up the powdery, reddish soil that covers most of Mars and blow it around.

JUPITER

At the top of Jupiter's colorful clouds, the temperature is a frigid -160°C. As you parachute down through the clouds, the air gets warmer, thicker, and cloudier. But don't expect to land soon. Jupiter's clouds are hundreds of kilometres high!

Jupiter's strong winds stretch its high-altitude clouds into long, colorful bands that curl around the planet.

Earth is like a big round rock that is covered by a thin layer of gases — the atmosphere. Jupiter is more like a giant globe of gas that has a small rocky core. Its atmosphere is six times as thick as Earth's.

There are no seasons during the 12 Earth-years Jupiter takes to orbit the sun. But there's always lots of weather. In fact, Jupiter's Great Red Spot is a huge storm that's lasted for over 300 years and is still going strong!

Jupiter's Great Red Spot is the largest storm in the solar system — it's wider than three planet Earths put side by side.

SATURN

As you approach Saturn, you'll see that its beautiful rings are not the solid bands they appear to be from Earth. They're really millions of ice particles that orbit the planet like tiny moons.

Like Jupiter, Saturn is a giant planet of gas. Since it's farther from the sun, Saturn is colder and lacks Jupiter's colorful cloud belts.

Unlike Jupiter, Saturn is tilted. So it has seasons like Earth and Mars do. But Saturn takes almost 30 Earth-years to orbit the sun, so its summers come only every 30 years!

Saturn's rings block sunlight, so parts of the planet are always in darkness.

Every Saturn summer, when the planet's northern part receives lots of heat from the sun, astronomers have seen a large storm erupt. These brilliant storms are caused by clouds of white ice particles that billow up from the depths of Saturn's cold atmosphere.

URANUS

Beyond Saturn, lies Uranus. As you float through the gas that makes up most of this planet, you won't see any spots, storms, or bands of clouds. But Uranus does have unusual seasons.

Uranus tilts so much that it spins on its side. It's as if it rolls around the solar system. During its 84-year-orbit of the sun (84 Earth-years, that is), the north pole of Uranus points straight at the sun for about 20 years. After that, Uranus's equator points at the sun for 20 years, then the south pole points at the sun for the following 20 years, and so on.

If this happened on Earth, our polar ice caps would completely melt every summer, refreeze in winter, remelt, and so on. We would experience much more drastic swings in temperature from summer to winter than we do now. Let's head back to Earth before we get fried — or frozen!

Astronomers are still not sure what effect Uranus's topsy turvy seasons has on its weather.

Zip, Snap, & wRap!

WHAT'S AHEAD

NATURE'S
Giftwrap

BY JAY INGRAM

**This is a tale
of how a package
can become more
important than
what's inside of it.
It's a tale of
the balloon fly.**

The story starts with the age-old question of where to find a mate. A long time ago, male balloon flies offered a present to any female who would choose them. They started bringing freshly killed insect prey when they came calling. This might have been to attract the female, or it might even have served to distract her from eating the male himself. Some insects do that. Whatever the reason, there are still balloon flies today that present female balloon flies with a juicy insect. Insect experts think these flies are species who just stuck with the original idea of giving an unwrapped gift.

A balloon fly

52

Other insects make packages, too. This is the spittlebug's foamy wrapping.

53

It's also possible to find some kinds of balloon flies today that prove that some males weren't happy with a plain gift. They began to decorate the insect with little bits of silken threads and little blobs of gluey stuff. This is not elegant gift wrap, but it is biodegradable and perfectly recyclable. The threads and globs may be there just to make sure that the insect doesn't struggle and get away. Regardless, the wrapping certainly doesn't bother the female: she eats the insect anyway.

These packages belong to the Japanese silk moth (top) and the giant silkworm moth (bottom).

Later, a few males discovered that the way to really show off the dead insects, and themselves, was to improve the packaging.

They wove the silk fibers in a spiral pattern to make a balloon-shaped ball with an opening at the back, a little like a globe with the Antarctic cut out of it. A beautiful glistening white ball —

with the dead insect plastered onto the front. Some went even further. They made the balloon much bigger, and stuck only pieces of the once-prized dead insect on the front, like sticking school pictures and coupons on your fridge with magnets. There are still male balloon flies today who woo their mates by making up these special dead-insect balloons. But the best is yet to come.

Can you guess the next step? That's right: a balloon, a very beautiful big balloon, with absolutely nothing on it — or in it. No old dried-up pieces of some dead insect stuck on the front. All wrapping — no present. The male flies who do this have achieved the ultimate: a package so attractive you don't have to put anything in it. Once male balloon flies had to be mighty hunters to impress the female with something particularly juicy and mouth-watering. Now, they have to be artists.

FLASHBACK

The Can, Can

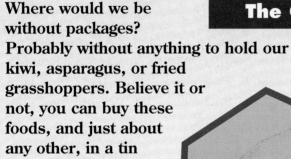

Where would we be without packages? Probably without anything to hold our kiwi, asparagus, or fried grasshoppers. Believe it or not, you can buy these foods, and just about any other, in a tin can, thanks to the work of Nicolas Appert. He was a French chef and pastry maker in the late 1700s. He was always interested in finding new ways to preserve food.

At the time when Appert was experimenting, France had an important reason for finding ways to preserve food. The country was going to war and needed to feed the soldiers. Appert developed a process that involved placing the food in glass containers, corking them, then heating them twice to remove the air.

Although Appert didn't know it, the heat was also killing microbes in the food. If the microbes weren't killed, they would make the food spoil and probably make people sick.

Appert won a large prize from the French emperor Napoleon for inventing a way to preserve food. He used the prize money to open the first commercial cannery in the world, the House of Appert. Although Appert died in 1841, his cannery operated until early this century.

Canned food today — both homemade and store-bought — is still made using almost the same method that Appert invented.

Right Off the

Did you drink milk or juice from a drink box today? Even if you didn't, you probably see drink boxes all the time.

Have you ever wondered how a drink box works? Just think — it holds your milk, but you don't have to put it in the fridge. Your milk at home has to stay in the fridge. What's the difference?

BY MICHAEL A. DISPEZIO

Milk in a bag, container, or carton has to stay cold, so it stays in the fridge. If the milk gets too warm, microbes can grow, and the milk will start to spoil.

But a drink box doesn't work that way.

A drink box is a microbe-free package. Inside it is a product (your milk or juice) that has been treated using the ultra-high temperature (UHT) process. In the UHT process, the drink is heated to about 150°C for about 10 seconds, then cooled. That destroys microbes. So drink boxes that aren't open do not have to be refrigerated.

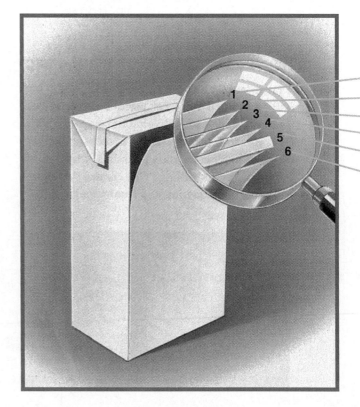

Shelf

Drink boxes are made from layers of paper, plastic, and aluminum foil. You can take a look at those layers if you pull apart an empty drink box.

The three materials are put on top of each other in six layers. Then they're squished together to make a single sheet. All six layers play a role in keeping everything from fruit juice to chocolate milk fresh and tasty.

The heavy paper, almost like cardboard, makes the carton stiff and strong. The carton's inner layer of plastic has two jobs. It doesn't react with chemicals, such as the acid in juices, that might change the taste of the drink. It is also waterproof. The carton's outer coating of plastic also has two jobs. It keeps the paper layer dry, and its clear surface is a good place for printing information about the drink.

Plastic

Plastic

Aluminum foil

Plastic

Paper

Plastic

These are the layers you will see if you peel apart a drink box.

Try THIS

Take a close look at a tin can. Remove the label and get an adult to help you remove both ends. See if you can figure out how the can was put together.

CHECK IT OUT!

Many kinds of packages are recycled. Find out from a local recycling group what kinds of packages are recycled in your community. What packages will you be able to recycle in the next year? In the next five years?

Aluminum foil in the carton stops oxygen, other gases, and microbes from getting in. If they were let in, they would start breaking down and spoil the drink. The foil also keeps out ultraviolet light. Some nutrients, such as riboflavin, are sensitive to ultraviolet light, and would start to break down if it got through.

With all of these layers you would think that drink boxes would be hard to recycle or reuse. In fact, companies in Europe and North America have found a terrific use for them. They collect empty drink boxes and turn them into tough plastic wood. The wood is used to make road signs, park benches, and even playground equipment!

THROWING IT OUT

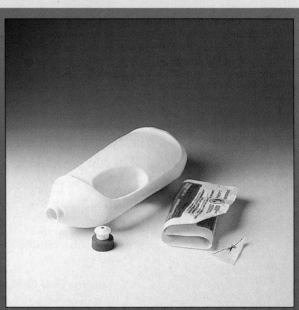

It's easy to see that an empty Enviro-Pak takes up less space than an empty bottle.

About one-third of your garbage is packaging that gets thrown out right away.

What are we doing about it? People are going back to buying bulk foods, making their own cleaners, and reusing bags. Many manufacturers are making packages that are smaller, lighter, and easier to recycle.

For example, companies that make

products like detergents and cleaners came up with the Enviro-Pak.

It's a soft plastic pouch filled with (for instance) liquid cleaner. You pour the cleaner either into a jar or into an empty jug that used to hold the cleaner. Instead of a plastic bottle, you only have a flat piece of plastic to throw away. It's only about as thick as two sheets of construction paper, and not quite as big as this page.

Some old movies feature the actors dancing on the ceiling and having a supertime.

Sound impossible? Well, it is for us. Those movies are just using trick photography.

But animals all around us can stick to walls, ceilings, food, and even you!

Turn the page to see how animals get a grip on the world.

A STICKY Situation

Can you hang by your toes from a smooth pane of glass? It's no problem for this lizard, called a gecko. If a surface is rough, a gecko uses its sharp claws to get a good grip. But if the surface is smooth, it uses its amazing toe pads. On each pad are rows of ridges. And on each ridge are millions of tiny brushes, each with about 2000 bristles. When a gecko walks, it pushes its toes down against the smooth surface. The tiny ends of the bristles poke into the smallest pits and cracks. Even a surface as smooth as glass has them. The gecko's bristles hold tight — until it pulls its toes free.

These feather duster worms are ocean-dwelling cousins of the ordinary earthworm. Feather dusters live in leathery tubes that attach to rocks under the sea. If danger threatens, the worm hides inside its tube. When all is calm, it pokes out feathery tentacles that are covered in sticky slime. Tiny plants and animals are snagged in the slime and swished to the middle of the tentacle crown — right into the worm's mouth.

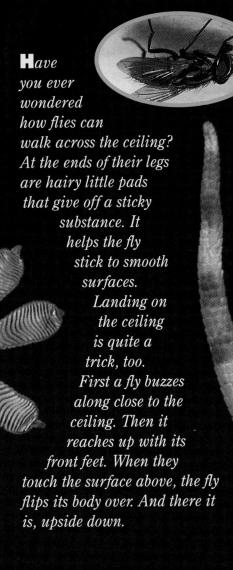

Have you ever wondered how flies can walk across the ceiling? At the ends of their legs are hairy little pads that give off a sticky substance. It helps the fly stick to smooth surfaces. Landing on the ceiling is quite a trick, too. First a fly buzzes along close to the ceiling. Then it reaches up with its front feet. When they touch the surface above, the fly flips its body over. And there it is, upside down.

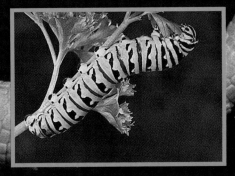

Caterpillars have tiny hooks on their feet so they can walk on any surface — including you. The little hooks catch on your skin. If you want to get the caterpillar off, you can hold a stick near it and let it crawl off you onto the stick. You can also use a tissue to pull it gently off you.

CUTTING EDGE

STRONG MUSSELS

Sticky stuff in nature often finds its way into human technology. Mussels, for instance, turn out to have, well, muscle. They are strong enough to cling tightly to the ocean rocks where they live.

Mussels are shellfish that live together in colonies. They produce a kind of glue that lets them remain stuck to the ocean rocks they call home. The "glue" these shellfish make turns out to be strong stuff.

Mussels stuck to their rocky home

A biochemist has discovered that the key ingredient is a certain kind of protein. Now manufacturers are working on a human-made version of the glue. What's the big deal? Most glues stick only to dry surfaces. This one is fast-acting and sticks to almost anything — even underwater.

Doctors could use the new glue to patch up broken bones, tendons, and even skin. Sailors could use it to do repairs on the bottoms of ships at sea. The possibilities are endless!

KEEP It Together

BY SUSAN PETERSIEL BERG

Just how do astronauts keep their food trays on their laps in space? How does a doctor close up a cut? And how do you keep your school reports together? Take a close look at these close-ups and see if you can identify the fasteners — and the neat things they can do.

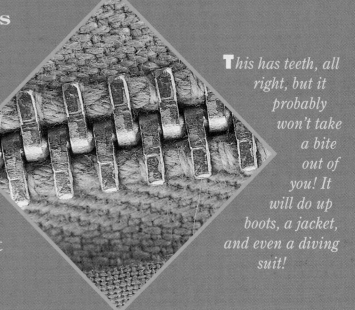

This has teeth, all right, but it probably won't take a bite out of you! It will do up boots, a jacket, and even a diving suit!

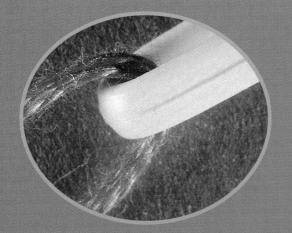

This used to be one of the only things around to fasten things together. But it's not old-fashioned. Tailors need this fastener every day — and so do doctors. It's something doctors use to finish up operations.

Believe it or not, sometimes doctors use these to close up the skin after an operation. You'll see them in school, in the top left corner of your reports. They also come in handy for people refinishing furniture. With just a few of them and some fabric you can cover an old chair and make it look brand new!

T*hey're stretchy, they're snappy, and even though they're simple, they're used for all kinds of things. Orthodontists use the tiniest ones on braces to help straighten people's teeth. The largest ones keep two or three new cars bundled together when they're shipped by train or boat from place to place.*

Y*ou probably think this sticky stuff is neat on your sneakers, but astronauts couldn't do without it. It helps them keep their feet on the ground and their food trays on their laps! What else do you think astronauts use it for?*

Answers: zipper, needle and thread, staples, rubber band, Velcro™

Try THIS

"How's business?"
"Sew-sew," said the tailor.

Write your own jokes about fasteners. Get a group of friends together and see how many jokes you can come up with. Be inventive!

CHECK IT OUT!

George de Mestral invented Velcro™. Look in the library for a book about inventors and find out how he got his idea.

Electrical engineers Michael Reed and Hongtao Han are working on a fastener whose microscopic hooks will hold blood vessels together.

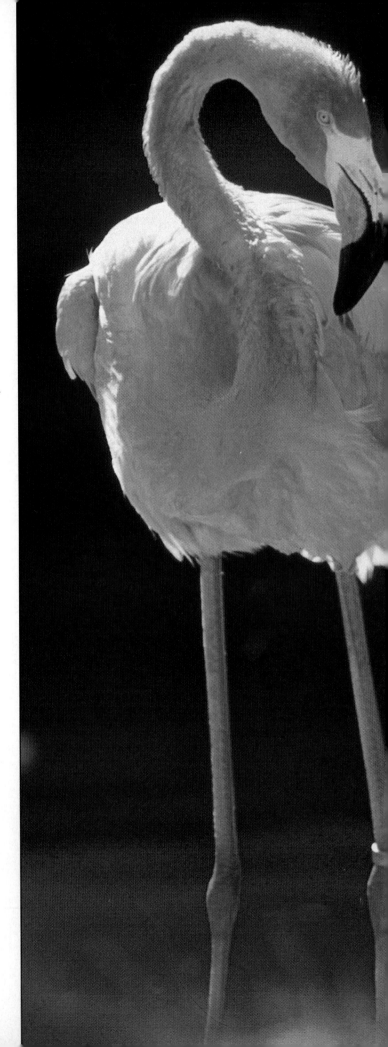

The Flying ARK

At the zoo, you're likely to see pandas from China, koalas from Australia, and lions and tigers from Africa. Have you ever wondered how those animals got there?

Well, for many years
animals had to travel
by ships and trains
and trucks.
It took a long time
to cross the ocean,
and some of the animals
got very seasick and
even fell and hurt
themselves in rough
weather.
Bumping along in trucks
or trains on long trips
could be dangerous, too.
By the time they reached
their new homes, many
were sick and hurt.

BY CAROLYN JACKSON

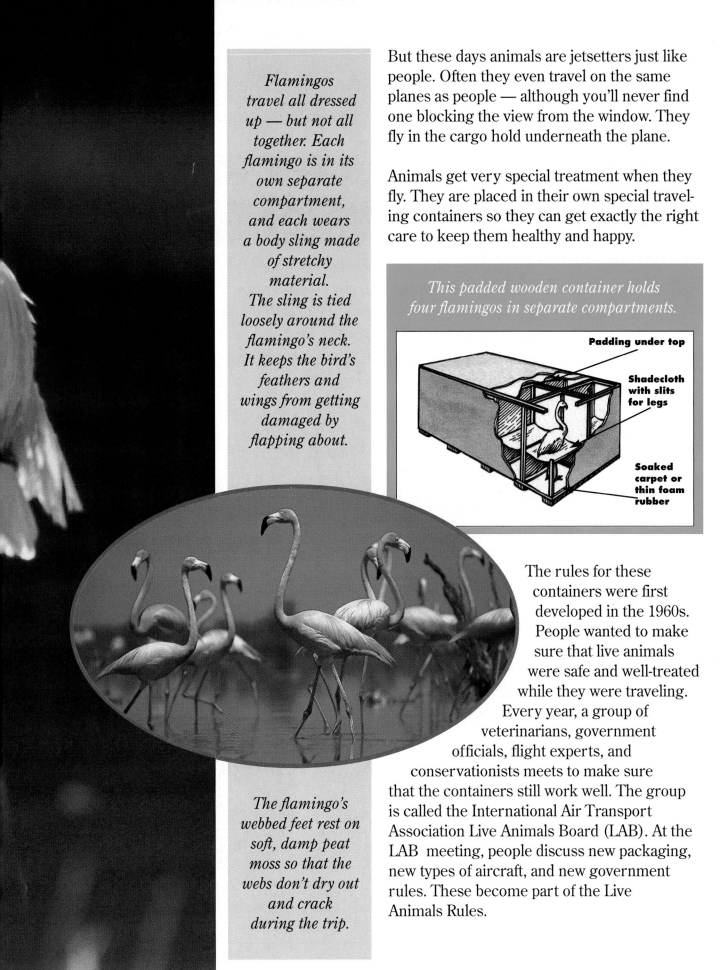

Flamingos travel all dressed up — but not all together. Each flamingo is in its own separate compartment, and each wears a body sling made of stretchy material. The sling is tied loosely around the flamingo's neck. It keeps the bird's feathers and wings from getting damaged by flapping about.

But these days animals are jetsetters just like people. Often they even travel on the same planes as people — although you'll never find one blocking the view from the window. They fly in the cargo hold underneath the plane.

Animals get very special treatment when they fly. They are placed in their own special traveling containers so they can get exactly the right care to keep them healthy and happy.

This padded wooden container holds four flamingos in separate compartments.

Padding under top

Shadecloth with slits for legs

Soaked carpet or thin foam rubber

The rules for these containers were first developed in the 1960s. People wanted to make sure that live animals were safe and well-treated while they were traveling. Every year, a group of veterinarians, government officials, flight experts, and conservationists meets to make sure that the containers still work well. The group is called the International Air Transport Association Live Animals Board (LAB). At the LAB meeting, people discuss new packaging, new types of aircraft, and new government rules. These become part of the Live Animals Rules.

The flamingo's webbed feet rest on soft, damp peat moss so that the webs don't dry out and crack during the trip.

Dolphins can leave the ocean and fly — that is, as long as they're in the proper package.

A dolphin doesn't have to be in water to breathe. It gets its air through a blowhole in the top of its head. Each dolphin travels in a kind of hammock made from canvas. Holes are cut out for its flippers to stick through and the whole thing rests on a thick foam-rubber pad. Someone must be with the dolphin at all times to spray its sensitive skin with water and to rub ointment on its head and flippers. It needs to stay moist just the way it is in the ocean!

Ointment on head

Sheet covering animal

Canvas stretcher

Foam rubber

Cutaway section for flippers

This waterproof box leaves room for the dolphin's head and tail.

Octopuses, fish, and some other water animals travel double-bagged in an insulated crate. The top of the bag is twisted, looped, and tied with an elastic band.

Top twisted and looped

Plastic

Elastic bands

Insulation at sides, top, and bottom

Let your pet travel with you! Design a package that will guarantee a safe, smooth ride. What materials will you use?

Design pet carriers for a car trip, a train trip, a boat trip, or an airplane trip.

An octopus likes to be in the water when it flies. Sounds kind of tricky on a plane, but it's not, really. An octopus sits inside a thick, strong plastic bag. The bag is less than half-filled with water and then oxygen is pumped in through the top before it is sealed. If the plane stops along the way, zoo keepers may come to the airport to give the octopus more oxygen.

A tiny computerized sensor can travel inside a package to collect information on the temperature, humidity, vibration, and shock a package goes through during its trip. If a package falls, the sensor can even tell how far it fell!

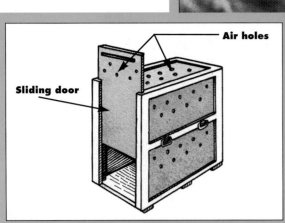

Air holes

Sliding door

Kangaroos have a special container, too. This container has lots of breathing holes. It's big enough so that a kangaroo can lie down and stand up straight in it.

And what does the Live Animals Board see for the future? Joseph Chan manages special cargoes, such as zoo animals, for the International Air Transport Association. He sits on the LAB and says that the board is working on new designs for bird containers. And in keeping with the worldwide idea of reducing, the LAB is hoping to cut down on the number of different kinds of containers it uses. It will do that by designing containers that will be the same for many different kinds of animals.

These special containers, and the work of the LAB, mean that animals are traveling — and arriving — safe and sound.

Kangaroos tend to get just a bit jumpy in a plane — and that can mean a big headache. So their traveling boxes have thick padding on the ceilings. That way kangaroos arrive feeling bouncy, but without lumpy heads.

THAT'S a WRAP!

BY CHRISTINE MCCLYMONT

Lunchtime! If you bring your lunch to school, you have to put it in something so you can carry it — maybe a cloth bag. And the pieces of your lunch? Well you have to put those in something, too. Your sandwich in a container to keep it fresh, your soup in a thermos to keep it hot, your cookies in a container to keep them whole.

This package was the creation of an unknown farmer in remote northern Japan. Her chickens were laying lots of eggs which she didn't want to break on her way to market. She also wanted customers to see that her eggs were the best. So she used the handiest material she could get — strong, flexible rice straw. She wove and tied the straws so that her eggs could be carried, stacked, and displayed, all without cracking!

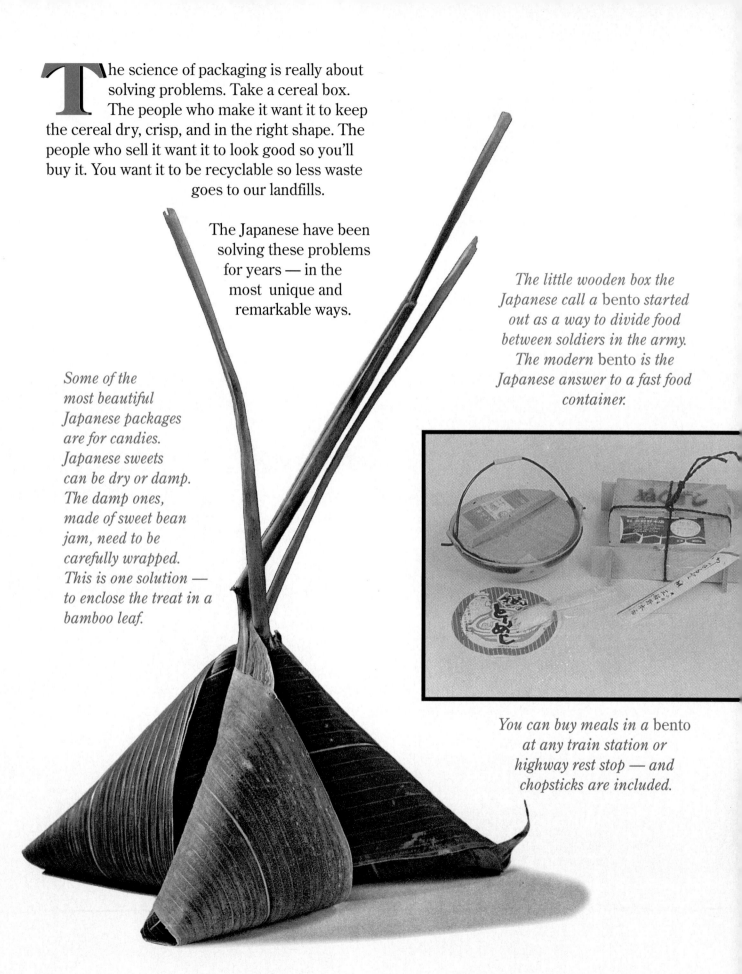

The science of packaging is really about solving problems. Take a cereal box. The people who make it want it to keep the cereal dry, crisp, and in the right shape. The people who sell it want it to look good so you'll buy it. You want it to be recyclable so less waste goes to our landfills.

The Japanese have been solving these problems for years — in the most unique and remarkable ways.

The little wooden box the Japanese call a bento *started out as a way to divide food between soldiers in the army. The modern* bento *is the Japanese answer to a fast food container.*

Some of the most beautiful Japanese packages are for candies. Japanese sweets can be dry or damp. The damp ones, made of sweet bean jam, need to be carefully wrapped. This is one solution — to enclose the treat in a bamboo leaf.

You can buy meals in a bento *at any train station or highway rest stop — and chopsticks are included.*

These items are all wrapped with a furoshiki (foo-ro-SHEE-kee). Its name means "bath cloth." When people went to the public bathhouse, it served as both a bathmat and a carry-all. The furoshiki folds flat to tuck into a bag or a pocket, so it's easy to carry. A carpenter might use a large, plain cotton cloth to carry tools. A visitor might choose a patterned silk cloth to deliver an elegant gift to a friend.

The beauty of a furoshiki is that it can wrap around an object of almost any shape. People use their cloths over and over again — a person might have the same furoshiki for a lifetime!

Be a packaging problem-solver. First collect found or recycled materials. How about last year's calendars, shoe boxes, dried flowers and leaves, yarn, and old clothes? Then think of some fresh new ways to package or gift wrap one of these items: apple pie, a bunch of carrots, spaghetti sauce, five eggs, theater tickets, a soccer ball, a book, or a pair of socks.

In 1600 the *bento* became a kind of fancy picnic basket. Families would have springtime parties to see the cherry blossoms and bring decorated wooden *bentos* with them. Today, there are as many as 1600 kinds of *bentos* found in Japan.

*It's fun to give gifts,
but in Japan it takes
a lot of thought, too.
It's not just what's inside
that matters:
the way you wrap
the gift should reflect
your feelings, too!*

*Origami,
or Japanese paper folding,
can be used to wrap
soft gifts such as scarves
and gloves without a box.
The beautiful origami
sculptures are as special
as the gift.*

So you've got a packaging problem? How to keep something warm, dry, moist, fresh, upright, flat, perfectly shaped? Solve it the way the Japanese would. Use good, available materials that can be reused. Build your package well and carefully, and think about the way it looks. When you're done, you'll have your problem all wrapped up!

BANANA BOX

Katsu Kimura is a Japanese designer who creates packages just for fun. This yellow box peels back like a banana. Inside there's nothing but another box! He has also made a square egg box with a crack that lets you see the square yellow yolk inside.

His Swiss cheese box is, naturally, full of holes. Kimura's boxes mimic the things they ought to contain.

Try it yourself! Design a funny package that mimics a doughnut, a hot dog, or your favorite food.

S p o t l i g h t ON

Light

WHAT'S AHEAD

Archimedes'
SECRET WEAPON

BY JAY INGRAM

You're a sailor on one of the most modern Roman ships in this, the Second Punic War. The war is against Rome's arch enemy Carthage, the mighty state on the north shore of Africa, right across the Mediterranean from Rome.

As the rowers pull your ship closer to the port of Syracuse, on the island of Sicily, your mind wanders. This has been a war featuring great personalities. The Carthaginians have Hannibal, the brilliant general who right now is roaming around in the country north of Rome, having brought his armies (complete with elephants!) through the mountainous Alps in the dead of winter. You're lucky enough to be sailing under the command of Claudius Marcellus. Better to be with him than against him. He's merciless to the enemy, but very good to his own. And you're sailing to a possible confrontation with the most mysterious and maybe the most powerful character in the war: the legendary Archimedes.

By all rights Archimedes shouldn't even be in this war. He's a great mathematician, not a general. Nonetheless he's turned his mathematical and scientific talents to building machines for the defense of Syracuse. You've heard horrific tales of arrows that seem to have been shot through walls, and giant cranes that reach out over the water, snag a Roman ship, and lift it right up in the air! And there are rumors that he has been working on something even more terrifying.

A snapped command interrupts your daydream. It's your job to help prepare the ladders, so that when your soldiers reach the city walls, they can climb over them before the Syracusians can react. That moment isn't far away now. The rowers have slowed their speed, and the ships are moving into place. A quick glance at the ramparts puzzles you: there seem to be so few soldiers there, waiting to stop your attack. Then ... FLASH ... a burning white light sweeps across the ship.

For seconds, you can't see a thing, but you can hear: panicked orders from the commander for the rowers to turn the ship and retreat, and shouts of "Fire! Fire!" As your sight returns, you can see fires burning everywhere on the deck. There's smoke everywhere, and as you rush to help extinguish the fires, you glance at the city walls. There, for an instant, you see a strange silvery object. As your ship begins to pick up speed, and the rest of your fleet, in disarray, begins to retreat, you can't help thinking .... was that Archimedes' latest weapon?

When the sun's rays bounce off a mirror, they send a lot of heat to one location and may start a fire.

Actually, that's a question scientists are still asking. Some ancient writers (writing a few hundred years after the end of the Second Punic War) claimed that Archimedes had

CHECK IT OUT!

Mirrors and light help get your message across in a fax machine. Light is shone onto the page being sent. Small mirrors help focus that light onto devices that turn the light into electronic signals. Those signals are transmitted to the receiving fax machine where they can be turned into words and pictures. Light and mirrors also work together in TV cameras and in telescopes. Go to the library and find out how these devices work.

But some people have actually tried to do what Archimedes did, and they've succeeded. In the 1700s a French scientist did experiments with a number of small mirrors, each of them only about the size of this book. Ninety-eight mirrors like this set a wooden plank on fire from 38 m away.

That may not sound like a great distance, but one of the ancient accounts suggests that the Roman ships were only a bow shot away when Archimedes set them on fire. That can be as little as 30 m away. Some people thought the ships might have been even closer to shore than that.

About 20 years ago there was more evidence: a Greek scientist, Ioannis Sakkas, lined up 60 Greek sailors, each holding a mirror, and got them to focus the sun on a wooden boat 48 m away. It caught fire almost immediately!

Carthaginian soldiers use small mirrors — lots of them — to reflect the light of the sun onto Roman ships to set them on fire.

Scientists have debated this question for centuries, but many scientists today doubt that it could have been done. It would have been difficult. Somehow mirrors on the shore would have had to be aimed exactly so that the sun's rays were focused onto a single ship long enough for the heat from the sun to ignite the wood. If the ship moved too quickly, or the aim was inaccurate, it wouldn't have worked.

But skeptics aren't swayed by these experiments. They think it's very doubtful that in the heat of war, with ships bobbing and swaying in the water, with arrows zinging through the air, that sailors would be able to aim mirrors steadily at a ship long enough for it to burst into flames.

What do you think?

RAINBO

Don't bother looking for the pot of gold at the end of this rainbow — or any rainbow, for that matter. Rainbows have no end, because they are circles. But there's another reason you can't see the end of a rainbow.

For you to see a rainbow, raindrops have to be falling somewhere in front of you, and the sun has to be somewhere behind you. That way the raindrops can reflect the light from the sun back to your eyes. You see that light as the colors of the rainbow because light from the sun is made up of all possible colors. Sunlight gets separated into those colors when it hits the raindrops.

If you try to walk to the end of the rainbow, the rainbow will keep moving with you. The drops are still in front of you, and the sun is still behind you, so you can never get to the end!

ALL LIT UP

Have you ever seen a prism (PRIZ-uhm)? It's a three-dimensional piece of glass or plastic that works just like a water droplet to separate the colors of light. Prisms usually have rectangular sides and triangular ends.

People hang prisms from fancy lights in their homes to fill their rooms with rainbows.

You can make your own rainbow — one that works exactly like one you might see in the sky.

On a bright, sunny day, take a garden hose and use your finger or a nozzle to spray a fine mist up into the air. You will see a rainbow form in the water droplets as they fall.

You can see the complete circle of a rainbow from an airplane.

79

A Light Touch

BY ERIC GRACE

Light is a form of energy.
Plants use light energy to
make their food.
A photographic film uses
light energy
to produce an image.
Scientists use light to do
many different jobs,
from making electricity
to sending signals
to helping diagnose
and treat various illnesses.

THE INVISIBLE SWITCH

Sliding doors seem to open like magic when you approach. The automatic doors are triggered at a light-sensitive photocell connected to the door switch. The switch is turned on by your shadow falling across the photocell.

LIGHT READING

Those little black bars that you see on every product may not mean much to you, but they say a lot to a beam of light at the checkout counter. When the cashier passes the bar code across the light beam, the beam "reads" the pattern of light and dark. It checks the pattern against a computerized list, then signals the cash register to print the correct name and price for the product.

LIGHT MUSIC

A compact disc player uses a beam of laser light in the same way that a record player uses a needle. The light beam reflects from the shiny metal surface in the grooves on the disc. Bumps in the grooves make a pattern of reflected light that is electronically converted into sounds.

LIGHT CONVERSATION

Little pulses of light traveling along thin glass fibers can be used to carry signals. These light tubes, shown on this page, are called optical fibers. They are being used in more and more telephone systems now. A single fiber, no thicker than a human hair, can carry more than 1000 telephone messages at the same time — and the fiber does this much faster than the usual electrical wires.

LIGHT POWER

When light strikes the surface of a solar cell, it produces electricity. This is known as the photoelectric effect. Solar cells are used instead of batteries in such things as pocket calculators and radios.

FLASHBACK

Ultraviolet light is important in medicine. Sister Judith Ward discovered that natural sunlight, which gives off ultraviolet light, can cure jaundice (JON-dis) in babies that are born early.

Jaundice makes skin turn yellow. It can be caused by liver or blood diseases. Many babies who are born early can't clean their blood properly, because their bodies aren't totally formed yet. So poisons stay in their blood and may cause blood diseases and jaundice.

In 1956 Ward was in charge of premature babies in a British hospital. When the weather was mild and

A Healing Light

sunny, she would take the babies outside. One day, when she was changing the diaper of a baby who had been outside, she saw something amazing. The skin that was still covered was yellow, but the skin that had been exposed to sunlight was a healthy pink. Doctors did experiments on blood samples from the babies and concluded that the sunlight had cured the babies.

Jaundice goes away by itself in most babies. But even today, babies with jaundice are placed under ultraviolet light. This helps bring down the poisons in the blood. It's a safe cure for a dangerous condition — thanks to Ward.

FOLLOW That BEAM!

A beam of light moves across a diamond, the hardest substance on Earth. Crack! The diamond splits. A beam of light strikes a thin steel needle. Zap! The light drills an eye, or a small hole, through the needle. A beam of light touches a wart. Sssssssssss. Heat from the light burns the wart off painlessly. The wart vanishes in a puff of smoke.

A beam of light that can do all these things, and many more, seems like a magic ray in a science fiction movie. This amazing tool is called a laser beam.

Laser beams have much more power than ordinary light beams. In ordinary light beams, tiny particles called photons (FO-tahns) travel on their own. They are out of step with each other. Scientists say that the photons are not coherent (ko-HEE-rent). When you turn on a flashlight, you may notice that the beam of light spreads, fades, and disappears a short distance from where you are standing. That happens because the paths of the photons quickly spread apart.

In laser beams, the photons all travel together, like marchers in a marching band. Photons in laser beams are coherent, so instead of spreading out they stay together.

Scientists produce laser light by using instruments that control the way the photons in the light beam move. Scientists use many kinds of materials to make laser beams. The beams may be different in color and have different characteristics, depending on the materials used. The power of a laser beam can be adjusted. Each laser beam is made especially for the job it has to do.

The list of jobs for lasers is long — and growing. Surgeons use laser light to make delicate cuts. The heat from the beam seals blood vessels as it cuts, making operations almost bloodless. Manufacturers use laser beams for cutting and welding metal.

Scientists use them to send signals into space and to measure distances to other planets. Lasers do a variety of everyday tasks, too. They can scan groceries for price codes at the checkout counter. They

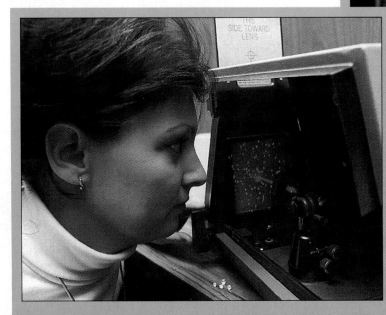

Jewel thieves, beware! Diamonds can now be "fingerprinted." A scientist beams laser light onto a gem through a hole in Polaroid film. The pattern in the diamond shows up on the film. Diamonds are crystals. Like all crystals, they form patterns. No two diamonds have exactly the same pattern, just as no two snow crystals do. The pattern of a diamond crystal cannot be seen with the unaided eye. But laser light reveals it and records it on film. The print can help a diamond owner identify and recover a stolen gem — even if it has been cut into a different shape.

measure air pollution. And they play the music on a compact disc in a compact disc player. That means that millions of people already have lasers in their homes!

84

This laser beam
cuts fabric
that is used
to make clothing.
It can cut
a perfectly
straight line
in no time!

The BETTER to See YOU with

Look!
Over there!
Can you see what's across
the classroom?
Some people
can see
what's across
the room
without
glasses,
but other
people can
see it only if
they're wearing
glasses.

To see clearly, your eyes must focus the light rays reflected by objects you look at. The rays enter each eye through a clear covering called the cornea (KOR-nee-uh). Then they pass through the pupil, the opening in the center of the colored iris, to the clear lens. All the parts are labeled on the diagrams on the next page.

Streetlights in San Diego are now required to wear shades. They were glowing too brightly for scientists at the nearby Mount Palomar Observatory to see the stars!

The cornea and lens focus the rays on the back wall of the eye, or the retina (RET-en-ah), forming a sharp image. If your eyeballs are too long, you're nearsighted. You can't focus well on distant objects. If your eyeballs are too short, you're farsighted. Nearby objects look blurry. If your cornea is not shaped perfectly, you will have astigmatism (uh-STIG-muh-tiz-uhm). This can add to problems in focusing on distant or nearby objects, or on both. Glasses or contact lenses correct the problems.

A person with normal vision would see the scene above. Both the rabbit in the foreground and the turtles in the background look clear.

A person who is nearsighted would see the rabbit clearly, but the turtles in the distance would be out of focus.

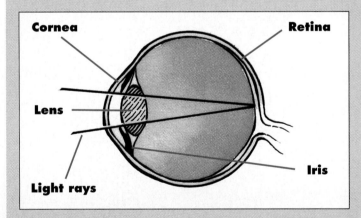

A normal eyeball is almost perfectly round. Light rays come to a point on the retina where the image forms. As this happens, both close and distant objects appear in sharp focus.

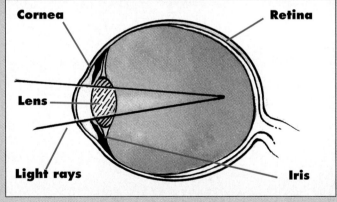

A nearsighted eye is longer than normal. Light rays from faraway objects come to a point and focus in front of the retina instead of on it. When the light rays do strike the retina, they form a blurred image. Only close objects can be seen clearly.

These photographs show you how the world would look if your vision was normal, near-sighted, or farsighted — and you don't even have to try on glasses!

To a farsighted person, the rabbit would appear as a blur. The turtles behind them would be in focus.

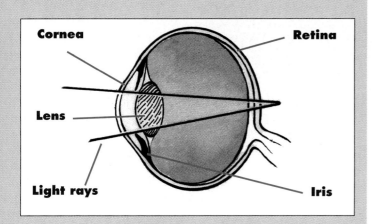

A farsighted eye is shorter than normal. Because of this, light rays reflected from nearby objects come together at a point behind the retina. This makes objects at close range look fuzzy. Objects farther back appear in focus.

LOOKING AHEAD

Can you imagine being able to see shapes and outlines through an artificial eye? Well, that's what researchers at Queen's University in Kingston, Ontario, are aiming for — and it may be possible.

Here's how the artificial eye would "see." A tiny television camera would be placed in a visually impaired person's eye cavity or on eyeglasses. It would take pictures and send them to a mini-computer in the brain. The computer would touch tiny electrodes in the person's brain, and create the shape of that picture.

Scientists are trying to find out how to create the picture. They have been able to attatch large electrodes to a patient's brain. So far they can get the patient to "see" phosphenes (FOS-feenz). They're sort of like what you see after a camera flash goes off — a bright spot on top of everything.

The next step will be to plant a tiny electrode in the brain of a visually impaired volunteer. Scientists will experiment with different ways of touching the electrode. That will help them find out the best way to make the volunteer see the phosphenes that create a certain picture. And that will help them know how to program the computer.

With the whole system in place, a visually impaired person would be able to see shapes and outlines, and move around a lot more easily.

MAGICAL BLACK LIGHT

At the Overbrook School for the Blind, in Philadelphia, Pennsylvania, excited children pour into the room. Someone turns off the regular lights and switches on the black lights. Suddenly, the children — all of whom are visually impaired — can see. Brightly glowing puppets and artwork seem to leap to life.

What causes this magic? It comes from specially colored toys and black light. Black light is a kind of invisible light, or energy. When it shines on certain substances, they absorb its energy and glow. The process is called fluorescence (flor-ESS-uhnts).

These glow-in-the-dark objects look about 20 times brighter than they would under normal light. With the help of black light, people who have very little vision can see many things.

Maria is surrounded by glow-in-the-dark objects. She has very little vision, but she can see these bright objects with the help of black light.

CHECK IT OUT!

Lots of things glow in the dark— including some kinds of food. If you chew on a wintergreen Lifesaver™ in the dark, other people will see what look like sparks coming out of your mouth! Talk to a science teacher or go to the library to see if you can find out why this happens.

Night Eyes

BY TONY SEDDON

Have you ever seen a cat's eyes shining in the dark? Drivers sometimes see a pair of shining eyes in the beam of their car headlights at night. This reflection of light from an animal's eyes is called eyeshine. It is found in some fish, certain toads, crocodiles, and snakes, and a few birds such as owls and the European nightjar.

It is most common in animals that are active at night.

Compared to its body size, this tarsier has the biggest eyes of any animal. On the same scale, each human eye would be nearly two metres across. The tarsier's pupils are so big that they cannot open up any more. Imagine the huge lens behind each pupil! These are a perfect pair of eyes for catching every bit of light on a dark night.

You might be surprised to learn that more animals come out to feed at night than during the day. This is because it is probably safer after dark. But there's not much light. Because of this, these night, or nocturnal, animals need a special kind of eye. Night eyes must be big with a large pupil and a huge lens. Eyes built like this can collect every bit of light available. Most night animals probably see four or five times better than humans can in the dark.

Animals and humans both have a layer of tiny sense cells at the back of their eyes.

There are two types of sense cells, called rods and cones. Cones are used for seeing in bright light and detecting colors. Rods respond to dim light. Nocturnal animals have many more rods than cones in each eye. This is why they can see better in the dark. But they see a shadowy, gray world. They don't see any color. Our eyes have more cones than rods. We can see colors, and we can see fairly well in bright and dim light.

But we still don't have eyes that are anywhere near as good at night as the ones on these animals — take a look!

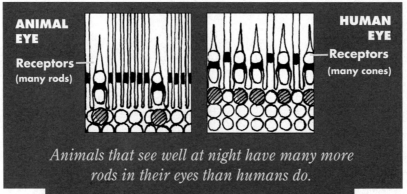

ANIMAL EYE
Receptors (many rods)

HUMAN EYE
Receptors (many cones)

Animals that see well at night have many more rods in their eyes than humans do.

This European nightjar has big eyes for seeing at night. A bright pair of eyes would give away its resting place, so it keeps them closed until dark.

In Paynes Prairie State Preserve near Gainesville, Florida, a photographer's flash picks up the light reflected by alligators' eyes.

92

MISSION: SURVIVAL

Puzzle

By Jay Ingram

Tyrannosaurus rex is the fiercest dinosaur that ever lived. Everything about this animal was horrifying. It weighed about nine tonnes and was 14 m long. That means it was about the size of a bus, and this bus had a mouthful of teeth like giant steak knives. The nearest thing to a T rex we have today is the great white shark. And on land, there's nothing that comes even close.

94

Not all dinosaurs were huge. A dinosaur called *Mussaurus*, which means "mouse lizard," was small enough to sit in your hand!

If your only impression of T rex comes from seeing cheap horror movies in which monsters like Godzilla rip high-rise office towers apart, you might be a little unsettled to realize that the real thing was much more threatening. The fake Tyrannosaurs in the movies waddle around like the clumsy models they are. But the real T rex was faster, stronger, and hungrier than any animal you've ever even dreamed of.

But while we know they must have been pretty fearsome, there's still a lot about Tyrannosaurs we don't know. All we have are their bones. From these, scientists try to paint pictures of what the whole animal was like. You should never forget that the real T rex became extinct 65 million years ago, before people even existed. So no one can ever be absolutely sure what it looked like or what it did. But there are some puzzling things about *Tyrannosaurus rex* that people are trying to figure out.

The legs of *Tyrannosaurus rex* are one of these puzzles. There's no real mystery about the hind legs — they were gigantic. They had to be, because the two of them supported all those thousands of kilograms of body weight. If you stood beside those legs, you wouldn't even come up to the knees! And those legs were powerful! Some scientists have estimated that Tyrannosaurs could have run as fast as 65 km/h in spurts.

On the other hand, the front-leg bones are very, very strange. They're only about 75 cm long. That's the same length as the arm of an average adult human. They look completely ridiculous attached to a 14-m, multitonne body. If Tyrannosaurs played cards, these "arms" would have been perfect for holding the cards close to their chests so the other players couldn't see them. But what were they really for?

Scientists disagree over whether the arms were long enough to reach the Tyrannosaur's mouth. But if they could, they would just barely reach, and probably couldn't have been much help in handling food while it was being eaten. Besides, there were only two claws on each arm. Surely they were there for a reason!

In 1970 a scientist came up with a bright new theory. This theory took into account that these arms were too short to be of much good for holding food,

Here's a T rex resting on the ground. Looking at its rib cage gives you an idea of how T rex lay on the ground.
(You won't see the rib cage in the following diagrams.)

T rex starts to get up by raising its head and straightening its "arms."

Next, T rex throws its head upwards and backwards and lowers its tail.

Finally, T rex straightens its legs. Now T rex's head is in its normal position and the animal is balanced.

or fighting, or anything like that. The scientist was intrigued by these tiny arm bones because, short as they were, they were very strong, and the bones supporting them were thick and strong, too. So they must have been used for something. The theory proposed that this enormous animal used its tiny arms to help it get up off the ground.

Picture a Tyrannosaur lying down, with its back legs folded under it like a chicken's, and its chin resting on the ground. How does it get up? Just dig in its hind claws and push? What you would have then is a huge *Tyrannosaurus rex* pushing itself along the ground, scraping its chin — it needs some leverage to get up.

The theory shows how the Tyrannosaur could dig its tiny claws into the ground and then push with its hind legs. The arms would anchor the animal and keep it from sliding forward, so pushing would lift its hindquarters into the air. Then all it had to do was throw its head back, pull its tail down, and stand up. It's funny, isn't it, that once you've pictured this in your mind, it makes perfect sense, even though no one has ever seen a Tyrannosaur do this and no one ever will. This picture comes from just a few bones and one scientist's imagination.

Of course, the idea that Tyrannosaurs did push-ups with their little arms isn't the last word. In 1990 some scientists looked very closely at some newly uncovered Tyrannosaur arm bones. They tried to get an idea of how big the muscles were that attached to those arms. They decided that, while T rex's arms might have been short, they were very strong. In fact, the animal's bicep (the muscle in your arm that you show off at the beach) was the size of an adult human thigh and could lift more than 180 kg.

What did it lift? There's another clue — it seems that the two claws on each arm face in opposite directions, like the barbs on a fish hook. Could these little arms have been used as hooks to hold the animal that T rex was eating after all? But there's still the problem of the short arms. The Tyrannosaur would have had to hug its prey to its chest to be able to use its arms.

There is one other possibility: maybe those arms weren't used for anything. As millions of years passed and dinosaurs changed gradually, these front legs got shorter and shorter. If the animal hadn't become extinct, it's just possible that T rex's front legs might have disappeared altogether — just as some scientists think we're gradually losing our baby toes. But it's not nearly as much fun saying the arms were useless as it is trying to figure out what they did do.

Do you have any good ideas?

AN OLD STORY

Have you ever heard of mass extinction? It happens when an entire population of animals dies out in a brief period of time. For years, scientists who study the past have known about such periods. The time when dinosaurs died out is one that still puzzles scientists.

Some scientists think we may be in a new period of mass extinctions today. It began about 11 000 years ago when ice-age glaciers retreated and the climate changed. Summers got hotter; winters, colder. The changes made life impossible for some plants and animals.

11 000 years ago The last ice age ended, humans migrated, and many large animals such as mastodons and saber-toothed cats disappeared. Were climate changes or human hunters to blame for their disappearance? No one knows for sure.

A.D. 500 — 1950 The arrival of humans on the islands of Hawaii, Madagascar, and New Zealand and in other isolated areas was a shock to resident creatures. The human hunters and their dogs, pigs, and rats quickly wiped out many animals such as the dodo, flightless ibis, and Steller's sea cow.

Present More than 1000 animal species are threatened with extinction. Even more plant species are endangered. Many people around the world are working to try to save these species.

FLASHBACK

A Living Fossil

You'd probably be stunned if you saw a dinosaur walk down your street. That's how scientists felt when a fish they thought had been extinct for millions of years was caught off the Comoros islands, near South Africa.

In 1938 a strange fish in a local catch caught the eye of Marjorie Courtenay-Latimer. She was a curator (that's a person who runs a museum) at a museum in South Africa. This fish was unlike any she had ever seen. Beautiful blue scales covered its long body, and it had unusual fins. Instead of lying flat, the fins sprouted out of the fish's body like paddles.

Courtenay-Latimer took the fish home to study it. Try as she might, she couldn't identify it. So she preserved the fish and sent it to a museum in England.

When scientists looked at it, they knew they had never seen a fish like this before, either. But they had studied lots of fossils of it. It was a coelacanth (SEE-la-kanth), a fish that scientists thought had been extinct for more than 70 million years!

Courtenay-Latimer had found a "living fossil." The first part of the coelacanth's scientific name, *Latimeria chalumnae*, was named after her in honor of her discovery.

Later when scientists investigated the Comoros islands, they found out that coelacanths had been turning up in local catches for hundreds of years. In fact, people had been using the coelacanth's unusual scales to patch bicycle tires!

Whooper

Ernie Kuyt, a biologist with the Canadian Wildlife Service, has an unusual job. Each spring, he flies around Wood Buffalo National Park in the Northwest Territories in a small plane, deliberately scaring whooping cranes off their nests.

K uyt's busiest time starts when the graceful, white cranes return in late April to their ancient nesting grounds in Wood Buffalo National Park. His first task is to discover and record how many breeding pairs of whoopers have survived the long migration from their winter feeding grounds in Texas.

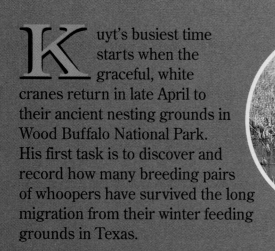

Now that one egg is tucked away in his sock and the other has been safely returned to the whoopers' nest, Kuyt must hurry back to the helicopter.

BY KATHERINE FARRIS

Two weeks later, Kuyt visits the nest sites again. This time he flies low enough to scare the cranes off their nests. He doesn't do this to be mean, he does it so he can count how many eggs each pair has laid. Whoopers usually lay two eggs, but they can only successfully raise one. And that's where Kuyt comes in. He can return later to each nest that has two eggs and remove one to be raised in captivity. But for now, he just counts the eggs and flies back to his base at Fort Smith. Once he's gone, the wary cranes return to their nests to incubate their eggs.

The pilot expertly brings the helicopter down close to the nest. Kuyt waits until the parent bird runs away, then he leaps out of the 'copter and splashes his way to the nest. But which egg to take? What if only one of them is alive? Quickly Kuyt gives each egg the water test. If the egg rocks or moves when he floats it, there's a live chick inside. Both eggs are alive, so he dries them, places one back on the nest, and slips the other inside a warm, woolly sock. He's back on board the helicopter in less than three minutes.

"We've found a nest with a bird on it. It's a white dot. Can you see it?"

About a week before the whooper eggs are due to hatch, Kuyt flies back by helicopter to all the two-egg nests. Now he's in a race against time. He has to collect all the extra eggs and get them into a warm incubator before the cold kills them. But he also has to work carefully so that he doesn't permanently frighten the adult whooping cranes off the nest.

As the helicopter heads out to the next nest, Kuyt notices with satisfaction that the parent bird is already stalking back to its nest. Another successful egg snatch!

The tiny beak breaking through the shell of the egg on the left leaves no doubt that this whooper egg has a live chick inside.

Once the eggs have been collected from all the other two-egg nests, Kuyt returns to Fort Smith with an armload of stuffed woolen socks. There, he gingerly places the large olive green and brown eggs in an incubator in his apartment. The next day the eggs are carefully crated and air-lifted to Maryland, where they will be hatched and kept in captivity.

In Maryland, a Canadian-American team of scientists studies the whoopers to learn more about them to help them survive. The scientists are working toward two goals. By the year 2000 they hope to have increased the number of whoopers thriving in the wild from 150 to 200. They also hope to set up a breeding colony with the captive whoopers by 2010.

By "stealing" one egg from each whooper's nest, Kuyt and the rest of the whooping crane rescue team have doubled the number of whoopers raised in a year. Whooping cranes aren't out of trouble yet, but their numbers have grown from an all-time low of 15 in 1941 to over 200 today.

CUTTING EDGE

ON THE RIGHT TRACK

What do you do when you're trying to locate and track an endangered species that roams widely over the ice caps? A Canadian company is hot on the trail of some new technology — and it's leading straight to the polar bear.

A large radio collar is put on a polar bear. The collar's transmitter receives signals from several satellites that orbit the Earth. They fix the location of the animal — within 100 m — and the collar stores this data. Scientists hop in a plane and follow the collar's radio signals. When they're close enough, they radio the collar to send the stored data to the airplane's receiver. Back at the office, the scientists load the information into a computer. That's how they find out where the polar bear roams!

— Susan Hughes

103

TO THE RESCUE

After scooping out a nest in the sand,
an olive ridley turtle lays her eggs
on a beach in Costa Rica.
Then she will cover up the nest
and return to the sea.
The eggs will hatch in 60 days.

Pushed by an inner drive to head for
the brightest horizon, newly hatched
leatherback turtles dash from their nest
toward the surf. Sea turtles usually hatch
at night, when any light from the sky
reflects on the water.

Some animals lay one or two eggs. But sea turtles are different. When she is laying eggs, a female sea turtle comes ashore every two weeks for a few months and lays 50 to 150 eggs. No wonder turtles have been around since the time of the dinosaurs!

But now turtle nesting sites are in danger. Animals and people are eating turtle eggs. Buildings line the turtles' nesting beaches. Oil and trash pollute their feeding grounds.

Scientists and volunteers are hard at work to help save those nesting beaches. They're even moving turtle eggs to safer spots.

They all hope to make sure that turtles swim Earth's oceans for centuries to come.

CHECK IT OUT!

Have you ever seen a wood turtle or a snapping turtle in the wild? These turtles are endangered in North America. Find out why.

Sharks lay the largest eggs in the world.

W*ild*1ife

WHO NEEDS IT?

What would happen if a Malayan pit viper

slithered into your kitchen?

Your dad would scream,

your grandmother would faint,

and you'd run

for your baseball bat,

right?

BY ANN LOVE AND JANE DRAKE

Before you finish off your unwelcome visitor, stop for a minute. There's something you should know about your intruder. The pit viper's venom makes a drug that helps prevent heart attacks in people.

Many wild animals and plants that you might consider creepy, ugly, or annoying are valuable. Like the pit viper, they provide much-needed medicines or food for people and other animals. And all plants and animals add to the variety and beauty of the Earth.

Malayan pit viper

The wild world is a rich and important natural pharmacy. Open your family's medicine chest and you'll probably see a bottle of aspirin or similar pain killer. Did you know that the ingredient that takes away headaches comes from the bark of a riverside willow tree? Aspirin can be made artificially using the chemicals found in the willow. But the heart drug digitalis (dij-ih-TAL-ihss) can be made only from the real plant, the foxglove. Thousands of heart patients owe their lives to this delicate plant. Many hundreds of wild plants and animals contribute important ingredients to medicines. Plants and animals save human lives!

The foxglove that you see here is used to make the heart drug digitalis.

Without wild plants we wouldn't have food to eat. Today's food crops are all related to wild plants. About 3000 plants are eaten by people, but only seven provide most of our food: wheat, rice, corn, potato, barley, sweet potato, and cassava (kuh-SAHV-uh). What would happen if one of these "big seven" was wiped out? Huge numbers of people might become sick or starve.

Willow tree

By protecting wild plants, we can keep our food crops healthy and preserve the world's future food supply.

Plants and animals also serve as food for other plants and animals. Mosquitoes might drive you nuts, but they're food for many amphibians and birds. Scientists see all living things as part of a large natural workshop called an ecosystem. Each

To prevent this from happening, farmers can crossbreed farm plants with their wild relatives that are immune to many diseases.

creature or plant, no matter how small or ordinary we think it is, plays a vital role in this ecosystem.

A WHALE OF A TIME!

After visiting the whales and dolphins at Sea World, 10-year-old Branwen Williams from southern Ontario decided she wanted to help them. For her birthday, Branwen adopted a whale. But she didn't stop there. After gathering all the information she could find, Branwen urged friends, family, schoolmates, and even grocery stores to buy only dolphin-safe tuna (tuna that's been caught with equipment that doesn't trap dolphins). "If everybody gets together, we can all save the whales and dolphins," says Branwen.

The loss of one plant or animal can have serious effects on others. Take the dodo tree and the dodo bird, for example. The dodo tree's fruit was the main source of food for the dodo bird. When the dodo bird became extinct the number of dodo trees dwindled. Then with only 13 trees left in the world, a scientist discovered why. The dodo tree depended on the dodo bird to eat the seeds in its fruit. The dodo bird's powerful digestive system cracked open the seed cases so that when the seeds came out in the dodo's droppings, they could sprout and form new dodo trees. The two kinds of dodos were interdependent; when the bird became extinct, the tree was threatened with extinction, too.

By 1681, about 100 years after the dodo was discovered, the last one had been clubbed to death. Until then people didn't realize that a species could become extinct.

Not all plants and animals are as useful as the pit viper, foxglove, and dodo bird. Some are just plain beautiful. People enjoy watching or photographing them. Wildlife enriches our lives and makes the world a more interesting place!

Try

THIS

Giant pandas depend on bamboo for food. Use these facts to calculate how much bamboo a giant panda eats in a week, a month, and a year. Calculate how much bamboo all the pandas that are alive in the world eat in a week, a month, and a year.

Giant Panda Facts

- About 800 to 1000 giant pandas are left in the world.

- Giant pandas live in bamboo groves in Southwest China.

- Every 50 to 100 years bamboo flowers and dies.

- A giant panda eats 35 kg of bamboo a day.

- Bamboo groves are disappearing.

MISSI

the Mystery of the Croaking Frogs

By Wendy Williams

The golden toads should have been jumping for joy. They had a clean mountain home in Costa Rica's Monteverde Cloud Forest Preserve, plenty to eat, and enough rain and mist to keep them moist.

cientists who visited the area in the early 1980s were thrilled to find hundreds of golden toads scattered around three different breeding sites. "They were like little jewels on the forest floor," says researcher Susan Jacobson of the University of Miami, in Florida.

But by 1989, the "little jewels" were gone. The golden toads had disappeared — maybe for good.

Costa Rica's golden toad

How do scientists know that amphibian populations are dropping? "We count frogs," says Dr. David Bradford, professor of environmental studies at the University of California at Los Angeles. "During the breeding season, we search shorelines and marshes for amphibians. We also listen for their calls at night."

Disappearing act

Fifteen years ago, yellow-legged frogs crowded the High Sierra Mountains in California. Now the frogs are gone from most of that mountain range.

No one knows for sure where the golden toads and the yellow-legged frogs went. But their disappearances are just two examples of a much larger problem. Throughout the world, many amphibians — frogs, salamanders, and toads — are disappearing fast.

The news has puzzled scientists. So they recently met to look into the problem. They found that some species are doing better than others. But overall, many amphibian populations are getting smaller.

Dr. James Vial is the director of the Environmental Research Laboratory in Corvallis, Oregon. He thinks there may be reason for worry. "Amphibian populations do go up and down," he said. "Species may disappear for one or two seasons, and then come back over the next few years. What bothers us is that the declines are so widespread."

A toad's heart is bigger than its brain.

Lately, there hasn't been much to hear. But scientists can't agree on any one reason for the silence. In Connecticut, scientists think it's the raccoons that are responsible for the decline of frogs. In Brazil, they think it's the cold winters, and in Canada, they think the problem is water pollution. No one knows why these amphibians are dying off.

SOS to the world

Amphibians are like living pollution testers. They can tell us a lot about the health of the world we live in. They are usually sensitive to a wide variety of environmental poisons. That's because they live part of their lives in water and part of their lives in damp soil. As a result, they risk being exposed to both land and water pollution.

Also, amphibians have no scales to protect them. They breathe in oxygen through their moist skin. And that skin is extremely sensitive to air pollutants and ultraviolet radiation — harmful rays from the sun.

What kind of an effect can air and water pollution have on amphibians? Dr. John Harte, a professor of energy and resources at the University of California in Berkeley, has studied tiger salamander eggs in the Colorado Rocky Mountains. The eggs had been exposed to acid snow. During the last seven years, he found that the tiger salamander population had decreased by more than half.

Bradford is not surprised. "Acid rain and acid snow are extremely poisonous to amphibian eggs," he said. "They also make adult amphibians get sick more often."

Stocking ponds and lakes with fish that eat tadpoles is another reason for the decline. Once the fish increase in numbers, the frog ponds become frogless. But the biggest problem may be the destruction of a frog's habitat — the area where it lives.

Okay, amphibians are disappearing. So what? Here's what: frogs, toads, and salamanders play an important part in the food chain. They eat insects and are themselves eaten by birds, mammals, and reptiles. Their extinction would throw nature out of whack. Many animals would be affected by this "missing link" in the food chain. And one day those animals, too, could disappear.

Tiger salamanders are missing in action.

Scientists don't know why amphibians are dying out. But because of the importance of amphibians in our world, they're determined to find out.

Going, going, gone

Janalee Caldwell and Laurie Vitt, two frog experts at UCLA, saw firsthand what happens when a frog's habitat is destroyed. They watched a huge logging operation in Brazil's Amazon rain forest. A stream of trucks took giant trees out of the forest. Then the settlers arrived. They cleared the land for farming by setting the whole area on fire.

"There are 75 species of frogs in that area," says Caldwell. "Very few of them would survive under those conditions. We're seeing habitats drastically reduced. You can't just keep destroying habitats and not expect things to disappear."

Paradise Lost

Earth's Rain Forests Are Disappearing

BY ELIZABETH VITTON

114

Zap! From below the surface, an archer fish blasts an unwary spider on a leaf with a bullet of water. The spider drops into the pool and is gobbled up. Above the jungle floor, a troop of monkeys chatter while colorful birds streak through the trees.

Welcome to the rain forest, an exotic world where the air is thick with mist, where fish walk on land, where passion flowers bloom, and tree frogs sing. A place where the constant hum of thousands of animals fills the air.

But now the roar of the chain saw is drowning out the jungle chorus. "Each minute, 40 hectares of the world's jungles are being cut and burned," says Dr. Stuart Strahl of Wildlife Conservation International. "At this rate, nearly all the Earth's tropical rain forests will either be destroyed or seriously damaged by the year 2035."

Competing for sunlight, trees in the tropical forest grow to amazing heights — 15 to 45 m — before sprouting branches and leaves. Below the tentlike canopy of trees grow palms, looping vines, orchids, ferns, and other plants. Less than two percent of the sunlight that nourishes the canopy ever reaches the floor. But even there, many plant and animal species thrive in a twilight world. Even

Most rain forests are found in the tropics bordering either side of the equator like a wide belt circling the Earth. Rain forests get an average of 250 to 1800 cm of rain per year.

It's a jungle out there! These "ant-sized" hikers walk on the wild side of a rain forest in Costa Rica.

At least 11 million hectares of rain forest are cut and burned every year. That means almost 30 000 hectares each day are lost forever!

though rain forests cover only a small area of the Earth, they are home to more than half of all living things on our planet. New species are being discovered all the time. Scientists think that for every species known, there remain 40 yet to be discovered. But nearly 10 000 species die out every year. Once a species is lost, it's gone forever.

One of the greatest dangers to animals is the destruction of their homes. List the ways that you could make your neighborhood or school yard a better place for wildlife to live. Think about things such as the use of pesticides and fertilizers, and the destruction of natural areas.

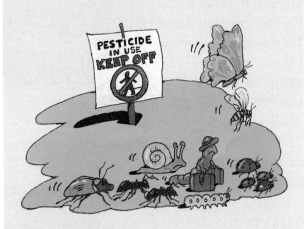

CHECK IT OUT!

Temperate rain forests run along the coasts of Alaska, British Columbia, Washington, Oregon, and northern California. These rain forests are in danger of disappearing, too. Find out more about them by checking in the library.

It's not easy being green

Why are we losing so much of our rain forests? Most of the tropical rain forests lie in poor nations whose populations are growing very quickly. Since they have no big industries, Strahl says, they make money by using the resources in the forest.

In just a few minutes a chain saw can topple a tree, but it will take hundreds of years for another tree to grow to the same size. The crashing timber destroys small trees lying in its path. Tractors flatten more forest when they drag the trunks to loading areas. As roads are cut to get the logs to market, it opens up the area to a flood of people who burn part of the jungle to make room for farms and ranches.

Destroying rain forests as far away as Brazil and Indonsesia has serious effects on all of us. For example, the world needs trees to recycle carbon dioxide, an odorless gas. Rain forests are the "lungs" of the planet. They suck the carbon dioxide out of the air through their leaves. The trees then "breathe" oxygen back into the air and pump it with moisture that falls as rain.

The rosy periwinkle is more than just a pretty jungle plant. It is made into a drug to help treat leukemia (loo-KEE-mee-ah).

Food for thought

Rain forests do more than help provide oxygen for the planet. They put medicine in your cabinet as well as food on your table. In fact, one-quarter of pre-scription drugs used in North America come from tropical forest plants. Each day people eat or drink something that comes from the jungle. Fruits, nuts, spices, coffee, sugar cane, cocoa, and even chickens come from rain forests.

Some jungle products have surprising uses, says Strahl. "Take M&Ms™. The candy doesn't melt in your hand because it's coated with a harmless wax, which comes from tropical rain forests."

What's being done to save the globe's greatest natural treasury? Some countries are setting aside nature reserves and planting new trees. In Brazil, a rubber tappers' union is fighting for large areas to be set aside just for rubber production and the collection of fruits and nuts.

Time will tell whether efforts like these succeed in saving the world's emerald forests. "I see it as a race against time," says Strahl. "We have the power to save or destroy the rain forests."

Demonstrators protest to save the last U.S. tropical rain forest, which is located in Hawaii.

MEET DR. GALDIKAS

Galdikas has spent more than 20 years studying and working with orangutans in Borneo.

"Wading all day up to my armpits in black swamp water wasn't too bad," says Birute Galdikas. "It was those leeches! They would crawl under our clothing, hide in our socks, and even fall out of our underwear!"

Galdikas is talking about the early days of her work with wild orangutans. In 1971 she arrived in the thick, steamy rain forests of Borneo, a long way away from her native Canada. She and her first husband lived in a bark hut, and their closest neighbors were six hours away by riverboat.

Now, more than 20 years later, her camp is a "neighborhood" itself. It is filled and busy with students, scientists — and dozens of "orphaned" orangutans. Orangutans are orphaned when they are caught and sold by poachers.

Galdikas works with these orangutans to get them ready to reenter the wild. Galdikas is now the world's leading expert on the red apes. To study a wild orangutan, Galdikas follows it all day, every day, for as long as it takes to get close. "After a while it gets used to you," she says, "and it doesn't even seem to care that you're there." Once a minute, she writes down what the ape is doing. These facts give an idea of an orangutan's way of life.

Working with the orphaned orangutans is very important to Galdikas. She and her staff return every one they can to the wild. Orangutans, like their gorilla and chimpanzee cousins, are becoming quite rare. Each one Galdikas saves helps keep the apes that much further from disappearing forever.

— Sallie Luther

Elephant

BY CATHERINE JACKSON

I lay wide awake in my tent.
It was my first night on safari in
Kenya. I was listening to the
sounds all around me when I
heard a lion call across the
plains. Then it sunk in;
I was really in Africa,
far away from my home in
Canada.

Encounters

Earlier in the evening, one of our guides had made us stop talking to listen to a low, coughing grunt. It had ended in a long, deep growl that made me shiver. "A lion," the guide had said. Now I could hear lions right across the river from our campsite! Only the canvas of my tent separated me from the wild.

My family and I began our safari in the Masai

Jackson holding a tortoise in the Masai Mara

Mara — Kenya's grassy, rolling plains. The Masai Mara is a vast, open space. There is so much to see at once that it's difficult at first to see the animals in it. One time a whole herd of elephants was grazing on a distant hill and I didn't even notice them! I thought they were just shadows or stones. So I had to learn how to look very carefully at details in the landscape. With our guides' help, I was soon spotting lions, cape buffalo, giraffes, tortoises, baboons, antelopes, and zebras. But the animals I loved the most were the elephants.

One morning we got up early to look for animals while the air was still cool and they were still active. We climbed into the Range Rovers, rugged jeeplike vehicles, and drove out. While we scanned the landscape from our perch on the Rover's roof, our guide spotted a large group of adult elephants and calves, feeding on a hill in the distance. As we drove toward the group, our noisy, smelly Range Rovers startled them. They trumpeted in alarm and the adults formed a tight knot around the young, as they quickly moved away. Although full-grown elephants weigh over 5000 kg, they can move fast!

When this group was at a safe distance, they stopped and waited for a signal from the lead female elephant, or matriarch. The matriarch gives orders in times of danger and guides all the herd's activities. Usually the matriarch's sisters, cousins, daughters, and all of their children make up a herd.

We stayed where we were for awhile to let them know we meant no harm. Suddenly, the group relaxed and began to graze again. The matriarch had realized we were not a threat and somehow she let the group know. We managed to creep closer and the elephants began

to move all around us. One even came right next to our Range Rover and started to eat by pulling up the grass with its trunk. I was amazed at how silently these huge creatures moved on their big, round feet. Unless a branch snapped underfoot, the only sound I heard was a low rumble of the elephants communicating with each other.

In the rain forest of Aberdaire national park, thick trees concealed elephants that were only a few metres away. The only way to spot them was to watch for slight movements in the brush. Here we spied three male elephants at a watering hole. Two were young and the other was old — he had very long tusks. Nowadays, it's unusual to see one with such long tusks.

For many years, people have hunted them for their valuable ivory tusks. Sometimes, poachers will kill a whole herd in just a few minutes to get their tusks. Poachers can sell two elephant tusks for the amount of money that an average Kenyan makes in one year.

When an elephant is two or three years old, its tusks start to grow and they continue to grow throughout its life. Elephants use their tusks to peel bark off trees, lift and move things, dig holes, and as weapons.

In 1989 Kenya's President Moi set fire to a mound of elephant tusks worth $3 million that had been confiscated from poachers. Although ivory trading was banned worldwide in 1990, poaching continues.

Bull elephants leave the herd when they are about 14 years old. They usually live alone and just visit herds to mate. Sometimes young bulls spend time together before they establish their own territories. Or, like the ones we saw, young bulls may stay with an older bull to learn about survival.

probably been poached. Elephants bear their young one at a time, several years apart, and mothers keep their calves by their sides at all times. It's unusual to see an adult caring for more than one calf. Poaching is very difficult to control outside the parks. Some poachers hunt the oldest and largest elephants first because they have the largest tusks. Quite often the young are orphaned.

One night we drove up a dry riverbed that was outside the national parks. Suddenly, we saw a young adult with two calves behind her. They had been coming down to the riverbed to dig for water. Caught in the glare of our headlights, the young adult froze. Then she trumpeted and ran. The two calves followed her into the trees where she was waiting to lead them away. We turned off our lights and listened to their retreat. The adult's distress had been heard by other elephants. Trumpeting and crashing boomed on both sides of the riverbed as they, too, disappeared into the darkness.

Unlike those we saw in the national parks, these elephants had been more than startled — they were afraid. Our guide thought that the mother of one of these calves had

As the trumpeting and crashing faded, I realized how lucky I was to have seen the elephants. I hope these beautiful, intelligent creatures survive so our future generations may be fascinated by them and learn from them, too.

An elephant's trunk can pick up anything from tree trunks to peanuts. It can also suck up enough water at one time to fill eight large milk cartons!

123

THE RHINO MAN

How far would you walk to save a rhinoceros? Michael Werikhe has walked thousands of kilometres to save the black rhinoceroses of his homeland, Kenya. To many people all over the world, he is known as the Rhino Man.

Werikhe has always been concerned about threatened wildlife. As a child, he brought home small, injured animals and cared for them until they could be returned to the wild.

In 1975 there were 10 000 black rhinos in Kenya. In 1980, just five years later, only 1500 were left. Werikhe was compelled to try to save them. He set out to walk 480 km across Kenya in 1982. Along the way, he told everyone he met how close the rhino was to extinction. He asked everyone to donate money, for every kilometre he walked, toward the rhino conservation effort.

Since then, he has raised millions of dollars walking across Uganda, Kenya, and Tanzania; across Europe from Italy to Great Britain; and across the entire United States. The United Nations Environmental Programme (UNEP) honored him as one of UNEP's Global 500 in 1989.

Despite Werikhe's efforts, there are only 400 to 500 black rhinos left in Kenya today. Many have been killed for their horns, which can be sold for a high price. But the Rhino Man plods on saying, "The rhino will live or die because of us."

All CHARGED Up!

BRAINS

BY JAY INGRAM

**Can you list everything
in your house that uses electricity?**

TORMS

The lamps, the television, the VCR, the microwave, the CD player, and the toaster all use electricity. So does everything else that's plugged into the wall sockets. All those sockets get their electricity from wires that come from the main electrical box, which in turn gets its electricity from power lines outside. Don't forget all the toys and cameras that run on batteries — they are electrical, too. That's about it, right? Well, not quite. Walk to the nearest mirror and look straight into it. You're looking at one of the most electrified things in your home: your head or, more precisely, the brain inside it!

127

That brain is stuffed with about a hundred billion brain cells, and each one is electrified. A single brain cell is very, very small and the amount of electricity running through it is extremely tiny. But when millions or even just thousands of them turn on all at once, there's so much electricity generated that special detectors on your scalp can measure it.

What turns brain cells on? Anytime you see, feel, hear, smell, or even just think of something, cells in some parts of your brain switch on. Scientists can learn a lot about our brains by recording those bursts of electricity. They try to figure out exactly where the bursts are and what triggered them.

Dr. Helen Neville is a Canadian scientist at the Salk Institute in California. She's trying to find out what happens in our brains when we read.

Helen Neville

This information may help people who are working with children who are learning to read. Neville has made some surprising discoveries by attaching electricity detectors, or electrodes, to people's scalps, then letting them read a sentence on a screen in front of them. Each word of the sentence

appears briefly and then it's replaced by the next word. She's found that the way the brain reacts depends on what kind of word a person is looking at.

If a person looks at a word that describes something real — something you can see, taste, or imagine like the words "tiger," "dance," or "cake" — there's a sudden flurry of electricity at the back of the brain. But something different happens if the word is one

Your brain uses the same amount of power as a 10-watt light bulb.

of the many little words in English that don't refer to any specific thing such as "if," "and," "beside," or "but." These words trigger a surge of electricity from a group of brain cells at the front of the brain, only on the left side.

Neville has discovered that it takes at least 10 years for the left front part of the brain to mature. These little words that the left front side of the brain seems to be interested in (even though they're small and unexciting to look at) are words that you need to use to make yourself understood. For instance, "The ice cream is beside your shoe," means something different from "The ice cream is in your shoe." "There's a tiger on that side of the river," means something a little different from "There's a tiger on this side of the river!" These words are like a carpenter's nails and screws. Buildings are made out of wood, bricks, metal, and glass, but without nails, screws, and bolts they would just fall apart. It's the same with sentences: they need special words to hold them together and make them understandable.

No one, not even Neville, is exactly sure what's going on in the brain when there are these sudden bursts of electricity. It's strange isn't it — we never think about what kind of words we're reading, we just read them. But our brains are noticing the different kinds of words — and zapping right along as we read!

KEEPING THE BEAT

Medics rush in with a machine the size of a personal computer. An electric disturbance in a woman's heart has started it fibrillating — that is, beating wildly and unevenly. The medics place two paddles on her chest and there's an electric surge. The defibrillator (dee-FIB-rih-LAYT-ur) has shocked her heart into beating regularly again.

But what if the team hadn't managed to get there in time? Imagine this: a small device the size of a pack of cards is placed in the woman's chest. It's battery-operated and attached by electrodes to her heart. Every time her heart begins to fibrillate, this defibrillator automatically jolts her with electricity.

Valluvan Jeevanandam, a New York cardiac surgeon, has already implanted these tiny, life-saving devices in 16 patients waiting for heart transplants. He knows the device will help their hearts keep a steady beat until donor hearts are available.

TAKING CHARGE

Has this girl seen a ghost? No. She's visiting the Ontario Science Centre and she has just touched the Van de Graff Generator. The generator looks like a big metal ball. It sits on top of a pillar which contains a vertical conveyor belt.

So how is the generator making her hair stand on end? Well, it all has to do with electric charges. Everyone's body has an equal number of positive and negative electric charges in it. The generator does too — when it's turned off. But when it's turned on, the conveyor belt works just like a conveyor belt at the grocery store — except instead of transporting potatoes, it moves negative charges away from the ball and down to the ground. This means that the ball ends up with fewer negative charges — so it "wants" more!

If you touch the ball, the ball will take negative charges from your body. You'll end up with more positive charges than negative ones — and whenever lots of positive charges are together they begin to push apart. Think about your hair. If the positive charges in your hair started to push apart, each of your hairs would try to get as far away as possible from the other hairs.

Your hair would stand on end. And that's exactly what's happening to this girl!

— Susan Hughes

Create a hair-raising experience with a couple of friends. Look for a plastic slide in your school yard or neighborhood park. Have one of your friends stand beneath the slide with his or her head within five centimetres of the plastic.

Get your other friend to watch what happens to the hair of your friend under the plastic as you slide down the slide. Switch places until you have all had a chance to see what happens.

131

The Elec

BY RON HIPSCHMAN

Lightning used to make me dive under the covers of my bed when I was a kid. As I got older, fear gave way (partly) to curiosity. Where does lightning come from? I had heard that lightning flashed when two clouds bumped together. But even as a kid that explanation seemed pretty silly. During most lightning storms the sky was completely covered by groups of clouds. I never saw lightning when there were only separate clouds in the sky.

ric **Sky**

To understand lightning, I had to learn about electricity. I found out that everything in the world contains electricity. Electrical charges are responsible for holding things together — and sometimes for pushing them apart.

Everything is made of atoms. Atoms are so tiny that you can't see them under an ordinary microscope. They contain two types of electrical charges. The minus, or negative, charges are electrons. The plus, or positive, charges are protons. Negative and positive charges are attracted to each other in the same way that opposite poles of magnets are. Just as like poles of magnets push away, or repel, each other, positive charges repel other positive charges, and negative charges repel other negative charges.

So what does this have to do with lightning? Well, a lightning bolt is simply a huge number of negative charges on the move! The negative charges travel from a cloud to the ground. As they do, they can heat the air to about 28 000°C. The superheated air expands quickly, causing a KA-BOOM — a shock wave that we call thunder.

What causes a lightning bolt? It all begins in a thunderstorm cloud. A cloud is full of positive and negative charges, but in a thundercloud, these charges become separated. Why? No one knows for sure. Somehow water drops in the cloud become negatively charged. Since the drops are heavier than the air around them, they fall to the bottom of the cloud. Positively charged atoms are left behind and warm air sweeps them up to the top of the cloud.

If you are caught outside during a thunderstorm, don't stand under a tree. The tree might act like a lightning rod. If you become part of the conducting path, you'll be electrocuted. When lightning shoots through a tree, the tree's sappy interior can heat up so that it boils. Then the tree could explode!

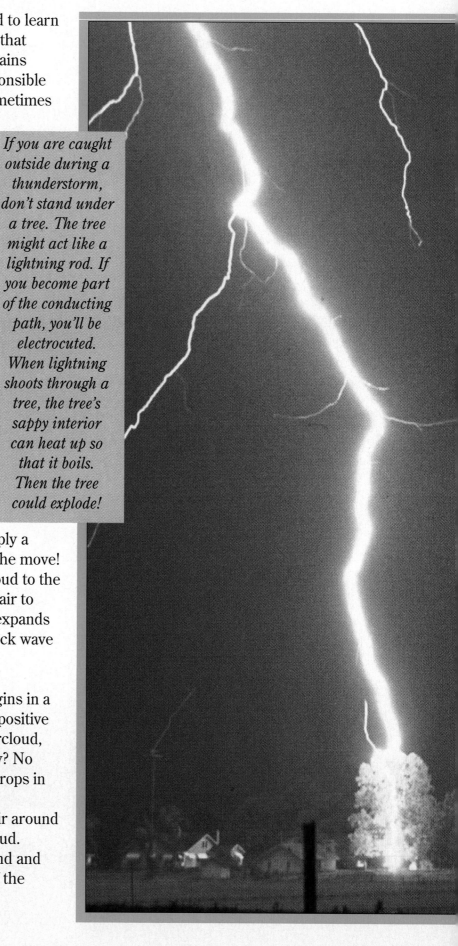

This sycamore tree was struck by lightning (left). Fortunately, it survived the strike unharmed (above).

Charge An amount of electricity. There are two kinds of electrical charges, called negative and positive.

Current A flow of electrical charges. A current can travel through anything that electricity can flow through — wire, metal, water, and even your body.

Electron Electrons are tiny particles that exist in all atoms. Electrons are much smaller than atoms, and 10 million atoms placed side by side measure only one millimetre. Electrons are negatively charged.

Proton Protons are tiny particles that are found at the center of atoms. Protons are about 1800 times heavier than electrons. They're positively charged the same amount that electrons are negatively charged.

In a piece of extension cord that's 10 cm long, there are more electrons than there are stars in our galaxy!

So the cloud now has positive charges on top and negative charges on the bottom — and the bottom of the cloud now has many more negative charges than the ground. The result? The cloud and the ground are attracted to each other, and the ground starts to pull the negative charges toward it. The charged particles act like a wire, making an electrical connection between the highly negative cloud and the positive ground. As the negative charges rush madly to the ground, they crash into the air. It is this collision that causes the "light" of the lightning. It makes the air glow with a bluish-white color.

Inside a thundercloud, electrical charges become separated. Warm air sweeps positive charges up, leaving the bottom of the cloud negatively charged. The attraction between the ground and the negative charges in the bottom of the cloud creates a lightning bolt.

Lightning strikes Toronto's CN Tower about 75 times a year. Tall objects are more likely to be struck because they bring the ground closer to the thunderstorm cloud.

air farther and farther up the "wire" starts to glow as the other electrons feel the electrical connection — and the bolt of light travels from the ground up to the cloud. Of course, this all happens extremely quickly — lightning flashes from the ground to the cloud in a speedy 98 150 km/s!

The safest place to be during a thunderstorm is inside a large building that has lightning rods. A car is also a good place because it surrounds you with metal. The metal will safely conduct the lightning's charge to the ground. Never stand under a tree!

If you could see it in slow-motion, you'd see that this lightning bolt starts near the ground because that's where the negative charges are connecting with the positive ground. Soon the

Now I know much more about lightning, and I'm still awed by it. I sit up and watch it streak through the sky at night, but sometimes I still feel the urge to dive under the covers.

BEN FRANKLIN, NASA STYLE

The first probe ever launched into a thunderstorm was Ben Franklin's homemade kite. He was convinced that the lightning in storms was the same as the electric sparks he generated in his laboratory. To test his theory, Franklin instruments on the ground measure the electrical charge in a storm cloud. When the charge grows very large, scientists launch a rocket. The rocket roars into the cloud, carrying a wire that brings positive charge from the

Blast off! Trailing a thin wire, a rocket heads into a thunderstorm.

A moment later, lightning strikes the rocket, instantly vaporizing the wire.

The wire glows and disintegrates, creating the trail to the right. Lightning, the jagged streak, then hits the lightning rod.

flew a silk kite with a key tied to its string near a thunderstorm in 1752. Electricity built up on the kite, traveled down the string, and struck the key. (Never try to repeat Franklin's experiment — you could be electrocuted.)

Today scientists in NASA's Rocket Triggered Lightning Program launch small, solid fuel rockets into thunderstorms. As a thunderstorm approaches, ground into the negatively charged cloud. This triggers a lightning bolt. The lightning strikes the rocket, follows the wire to hit a lightning rod near the launch pad, and its energy flows into a cluster of instruments.

By getting lightning to strike when and where they can measure it, scientists are looking for ways to protect people and objects from its destructive force.

137

THAT
Shocking
Eel!

BY ANITA GUSTAFSON

Our bodies are full of electricity.
We couldn't live without it!
But can you imagine
if we could channel that
electricity to our fingertips
and send out a zap
whenever we wanted to?

Sound impossible? Well, maybe for humans, but an electric eel can make and use electricity to protect itself. If a creepy creature gets too close, the eel can shock it by sending out a great jolt of electricity. Its electricity also helps the eel to find its way in the dark oceans and to find and catch its food.

The electric eel is one of the most dangerous animals in the rivers of South America. It can grow to be 1.5 m long. When a full-grown eel is disturbed, it can release electric shocks of up to 650 volts. That's enough to knock out a person!

Where does all this power come from? The eel makes electricity in special cells. These cells are grouped together to become the fish's electricity-making organs. These organs run up and down the eel's body. They're like the fish's "batteries."

The electricity organs in the eel's tail send out a weak electric current. It flows through the water and around to the eel's head. The current surrounds the eel, and when something touches the current, the eel can sense it.

If the eel senses a large animal that may be an enemy, or a small one that may be something to eat, it turns on its full power — a tremendous jolt of electricity. Frogs and small fish are killed or stunned by the jolt. Then the eel catches them easily. Enemies may be killed, knocked out, or frightened away.

Scientists don't know why, but eels don't seem to be able to shock other eels. And they don't know for sure why an eel's electricity doesn't shock the eel itself.

Scientists believe that it developed its own way of making electricity long, long ago. In fact, the electric eel was shocking its way through the water when dinosaurs roamed the land!

Sound fishy? It's all true!

CHECK IT OUT!

Deep-sea angler fish give off light like fireflies. This is called bioluminescence (BY-oh-loo-mih-NES-sens). Find out how these fish use bioluminescence.

The name Noctiluca means "night light." When millions of these tiny aquatic creatures get together, they give off enough light to read a newspaper by at night.

fernandez

OUT OF THE

If you fly over a city such as Montreal, Tokyo, Los Angeles, or Paris at night, sparkling lights will spring out of the darkness that spreads as far as you can see.

BY C.L. BOLTZ

DARKNESS

As your plane descends, you will be able to pick out white and orange street lights, endless streams of car headlights, the warm glow of houses and apartments, the slightly eerie brilliance of factories and office buildings, and the different colors of the airport lights — green and red and purple and orange.

Can you imagine how much electricity is needed to produce so much light and heat? Few of us ever pause to wonder where it all comes from or even how it is made.

All electricity is created by some kind of energy or movement. Here are three ways of creating electricity: by using the energy in fossil fuels, by using the energy in atoms, and by using the natural energy in water.

141

Electricity from fossil fuels

Fuels such as coal, gas, oil, and diesel are an important source of electricity today. They're called fossil fuels because they are the ancient remains of plants and animals that are found in the Earth. Fossil fuels have to be burned before they release the energy needed to make electricity. They're nonrenewable and pollute the air when they're burned, so we are always trying to come up with other ways to get the energy we need.

Electricity from atoms

Nuclear (NEW-klee-ur) power reactors produce steam to turn turbines and create electricity, just like fossil fuel furnaces. But they create the steam in a different way through a process called nuclear fission. In nuclear fission, the nuclei (NEW-klee-eye) of uranium atoms are split in two. This splitting of the atoms creates energy and heat.

Here's how fossil fuel furnaces and nuclear power reactors can provide the electricity to light up your bedside lamp.

From fossil fuels ...

1. **Burning Fuel** *In the boiler furnace a fossil fuel such as coal, oil, or gas is burned to create enormous heat.*

2. **Making Steam** *The heat created by burning the fuel raises the temperature of water as it circulates in the boiler-tubes and turns it into steam.*

3. **Spinning the Turbine** *The steam travels along pipes to the turbine. It hits the turbine blades and makes them spin.*

4. **Driving the Generator** *The spinning turbine-shaft drives the generator. The steam is turned back into water and recycled.*

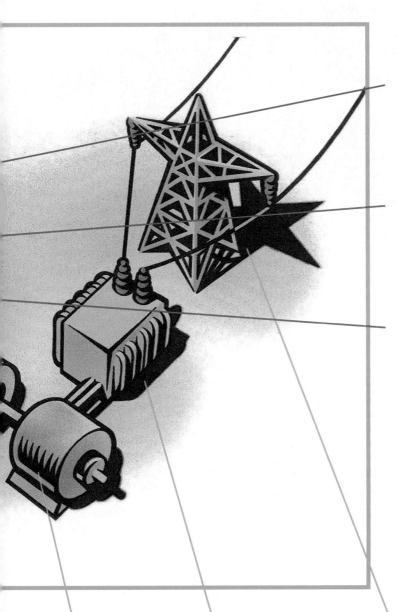

1. Heating Water *Nuclear fission heats this water to about 300°C. The water doesn't boil because it's under such high pressure.*

2. Making Steam *The water moves to a second water circuit. Its heat warms this second source of water. Because it is not under pressure, the water in the second water circuit boils and produces steam.*

3. Steam to Turbine *The steam travels along pipes to a turbine just as in a fossil fuel power station. After that all the steps are the same.*

The amount of electrical energy used by a 100-watt light bulb in 10 hours would be enough to boil eight litres of water or run a color TV for three hours. Lights out when you leave please!

7. Sending Electricity *The electricity flows into a grid system of overhead lines and underground cables. Transformers in other parts of the province or state change the high voltage to lower voltages for use in factories and homes.*

6. Transforming Electricity *The transformer changes the electricity from a low voltage to a very high voltage because this is the best way of transmitting electricity over long distances.*

5. Generating Electricity *The generator produces electricity, which flows through cables to a transformer.*

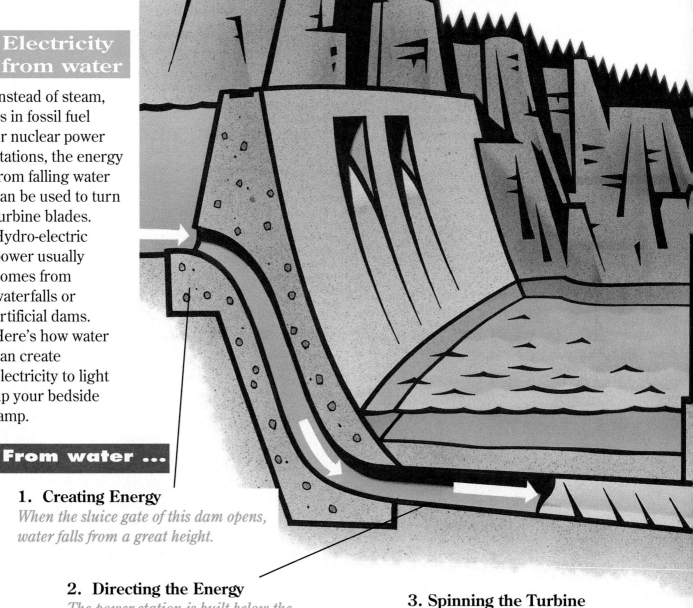

Electricity from water

Instead of steam, as in fossil fuel or nuclear power stations, the energy from falling water can be used to turn turbine blades. Hydro-electric power usually comes from waterfalls or artificial dams. Here's how water can create electricity to light up your bedside lamp.

From water ...

1. Creating Energy

When the sluice gate of this dam opens, water falls from a great height.

2. Directing the Energy

The power station is built below the dam. The water rushes along a pipe to a water turbine in the power station.

3. Spinning the Turbine

The moving water turns the turbine blades.

BLACKOUT!

Just as the evening rush hour began on November 9, 1965, the Niagara Falls electric power network broke down. Thirty million people in Quebec, Ontario, Pennsylvania, New York, Vermont, New Hampshire, Massachusetts, and Connecticut

5. Returning the Water

The water is then piped into the river.

4. Generating Electricity

A rotor attached above the blades spins and drives a generator which produces electricity. Then the electricity follows the same path as the electricity from fossil fuel or nuclear power plants.

Generate a little energy yourself and see if you can find out which one of these methods is lighting up your town — and that bedside lamp of yours!

suddenly found themselves in the dark.

You might have been in a blackout when a thunderstorm, blizzard, or hurricane put your power out. What's the first thing you do? What happens after a couple of hours? What happens after 12 hours? After 24 hours?

Try THIS

Find the electrical meter for your house or apartment. The wheel that spins inside it shows the amount of power your family is using right now. Turn off all of the electrical lights and appliances in your home and watch the wheel slow down.

Have a friend or family member turn on as many lights and appliances as possible and watch it go! The more energy you use, the faster the meter runs and the more money it costs your family — and the planet. Make a list of 10 things you can do to conserve energy.

CHECK IT OUT!

Does your family pay the hydro bill or the electric bill? Find out what the source of electricity is for your neighborhood. Is your school powered by the same source of electricity?

145

Conductor DEDUCTION

Y ou have snuck into a vault
in your mom's lab.
The vault holds rare species
of insects that you want to sketch
for your science project.
An hour ago, a security guard
came by and sealed the vault
for the night.
But you didn't panic, except for thinking
you might get caught.
You know there's an emergency
circuit to let you open the door
from the inside.
Now you've got all the sketches
you need, and you're dying
to get out of the cramped,
stuffy space.

BY SANDRA MARKLE

You open the panel of the emergency circuit and your mouth drops open. You can't believe it — about 15 cm of wire are missing! The circuit is broken. You're trapped. You groan, wondering how much air you've got left.

You bang on the door and call for help. There's no answer. You glance at your watch — oh no! you're late for dinner again. How are you going to get yourself out of this one? Suddenly, you get an idea and dump everything in your knapsack on the floor. Maybe you've got something you can use to replace the wire and complete the circuit.

Take a look at the illustration of the pile of things you have with you. What can you use to span the 15-cm gap to complete the broken circuit?

Check your solution on the next page.

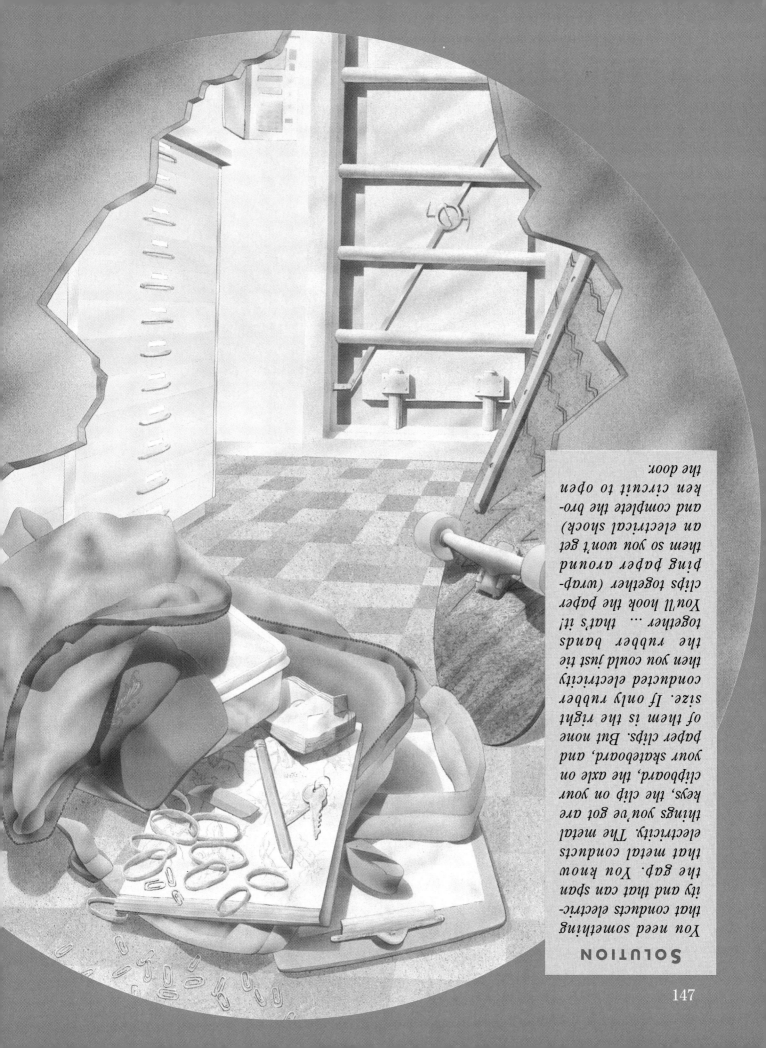

SOLUTION

You need something that conducts electricity and that can span the gap. You know that metal conducts electricity. The metal things you've got are keys, the clip on your clipboard, the axle on your skateboard, and paper clips. But none of them is the right size. If only rubber conducted electricity then you could just tie the rubber bands together ... that's it! You'll hook the paper clips together (wrapping paper around them so you won't get an electrical shock) and complete the broken circuit to open the door.

On *the* Fast

BY CHRISTINE MCCLYMONT

Trains are coming back in
style. But these trains don't
puff out smoke or burn
fossil fuels. These are the
trains of the future —
electric trains.

T r a c k

Believe it or not, electric trains are over 100 years old. Back then steam railways were all the rage. But the discovery of electricity caused lots of excitement. Inventors rushed to use the new power. It took a few years to get an electric motor working. But after that, electric trains were right on track.

How can you tell if a train is electric? A sure sign is a long line of poles that hold overhead wires above the railway tracks. Unless you've been to Europe or Japan, you may never have seen an electric train. But you may have ridden on a subway or streetcar. Subways run on electricity they pick up from a third rail that runs beside their tracks.

Streetcars are powered by electricity, too. They have an arm, or pantograph, on top of them that runs along an overhead wire picking up electricity.

Since electric trains were invented, Switzerland, Italy, and other European countries have electrified their railways. But Canada and the United States stuck with steam engines for a long time and then switched almost completely to diesel. Why?

The main reason is that North America is so big. Diesel trains can travel long distances carrying their own fuel, and they run along ordinary train tracks. Electric trains need a nonstop source of electric current. Setting up a network of poles and overhead wires all the way across the continent would be very expensive.

Should North American railways switch to electric trains? Many people say yes, for two reasons: the environment and speed. Although diesel trains pollute much less than cars and airplanes, the fossil fuel they burn still harms the environment. And if you've ever seen one pull out of a station, you know they're slow. Once they're going, they can cruise along at about 120 km/h, but they take a long time to stop.

Electric locomotives are quiet and the trains themselves don't produce smoke or exhaust fumes. No wonder they're popular in cities! Of course, the electricity that runs the trains might be generated at a power plant that uses fossil fuels — but it takes a lot less fossil fuel to run an electric train than a diesel train, a bus, a car, or an airplane. Although the power plant creates some pollution, it's not much compared to the amount of pollution created by other kinds of transportation. Just think — an airplane creates about 50 times more pollution than an electric train. Electric trains are also more powerful than diesel engines. Because they can draw huge amounts of power from the central power plant, they can whiz along at high speeds. When trains are fast and reliable, more people use them. So electric trains make good sense on heavily traveled lines between big cities.

The Shinkansen, Japan's high-speed electric train, zips along mountain valleys, through tunnels, and over long bridges between Japan's major islands at speeds over 200 km/h. It runs on its own elevated track, picking up power from an overhead wire. A main computer in Tokyo controls over 150 trains as they speed in and out of major cities every few minutes. And they're always on time!

CHECK IT OUT!

Have you ever heard of a "maglev"? It's a high-speed train that "floats" on a cushion of air as it zooms along. Take a trip to the library to find out how these trains float and where they're used. Share the information with a friend!

150

The new generation of electric trains is changing everybody's idea of train travel. The main reason? Computers. These electric trains can be operated automatically. This means that trains can travel much faster and go in and out of stations much more frequently than diesel trains. They could even replace airplanes as the fastest way to travel between big cities!

Look for high-speed electric trains zooming between large North American cities in the twenty-first century. With cars and airplanes polluting our world, electric trains are the way to go.

France's electric TGV lives up to its name. TGV stands for train à grande vitesse *which means "high speed train." The TGV usually travels at over 250 km/h and can go as fast as 300 km/h. A special version has actually been clocked in at 515 km/h! Although the TGV runs almost on "automatic pilot," it still has a driver to keep an eye open for anything unusual.*

Instead of a subway, Vancouver has a "skyway." It runs on special elevated guideways and is powered by electric motors that fit beneath each car. Don't be surprised if it pulls into the station and you discover there's no driver. The SkyTrain is fully computerized. The fastest it can go is about 80 km/h. But it's not designed to break speed records.

FLASHBACK

"Warning. Do not proceed. Broken rails ahead."

Train Talk

This is the kind of message that a train telegraph operator was lucky to get in the late 1800s. A message like this would save many lives.

But many important messages did not make it from the station to the train. There was a basic problem with train telegraphic equipment. A wire linked the telegraphic equipment in the train to the station. To receive or send a message part of the moving train had to touch the wire. But very often the contact between the train and wire was poor. Many messages were interrupted or incomplete and many safety warnings were never received. Often this resulted in serious train accidents.

It was Granville Woods who decided to do something about it. Woods had always had an interest in both railroads and electricity. When he was a boy in

Ohio, he read every book he could on electricity — and he never stopped thinking about ways to improve the control and distribution of electricity. In fact, by the time Woods began working on the "broken messages" telegraph problem, he had already been awarded patents for over 50 inventions, many of them to do with electricity!

So how did he try to solve the problem? Woods knew that two wires don't need to touch for electric currents to pass between them. Woods ran an electric wire down the track between the rails. He connected that wire to telegraphic equipment at the train station. He also ran an electric cable under the train's telegraph car and connected it to the train's telegraphic equipment. As the train moved, the cable traveled above the electric wire and electric currents could pass between them. So when messages were sent from the station to the train — or vice versa — they always reached their destination.

Chemical

COOKS

And You Thought **Cauliflower** Tasted Bad

BY JAY INGRAM

See if you can wrap your tongue around this name: denatonium saccharide (dee-nuh-TOE-nee-um SAK-uh-ride). The name is the only part of this chemical you'd want to wrap your tongue around, because denatonium saccharide is the worst-tasting stuff in the world. There's nothing that's more bitter than this chemical.

Denatonium saccharide (its commercial name is Vilex) came from another chemical that was put into paints and other dangerous liquids to keep people — especially young children — from drinking them. But the chemical wasn't bad enough.

It didn't keep rats from gnawing through cables, or other animals from eating plants. So chemists did some experiments and came up with Vilex. It's so bad that if just one drop was added to 3000 one-litre bottles of pop, it would make them all taste awful. That's just one drop of Vilex in 100 million drops of pop. How can something taste so bitter?

Your taste buds for bitterness are arranged across the back of your tongue. They're the most sensitive of all your taste buds, and that's just as well, because most poisons are bitter. Each taste bud contains little pockets called taste receptors which are made to catch certain kinds of chemicals.

If you were able to watch a receptor for bitter tastes in action, here's what you might see. (You have to use your imagination, because receptors and the food molecules they catch are much smaller than anything you can see in most microscopes.)

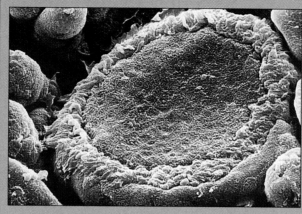

Taste bud

A single molecule of something very bitter enters your mouth. It bumps against the receptor and the two snap together. They connect a bit like a piece in a jigsaw puzzle, but with a much more complicated, and much, much smaller, three-dimensional shape. When the bitter chemical locks on, the receptor changes shape slightly and a nerve at the bottom of the receptor fires a message to the brain. That message tells you that you've got something terribly bitter in your mouth. Usually your mouth is full of the chemicals you're tasting, so probably thousands of messages are firing all at once.

But with denatonium saccharide you don't seem to need thousands of molecules to send the message to your brain. Just that one drop in 100 million does the job by itself. Scientists aren't sure why so little of it can taste so bad. Maybe it's that this particular molecule fits absolutely perfectly into bitter taste receptors, and just sits there. Other, weaker bitter substances may just touch down, then leave again.

Did you notice something funny about the name of the world's worst tasting substance? Its second name is saccharide. Ask any chemist, and you'll discover that a saccharide is a sugar. But sugar is sweet. How can it be bitter, too? Saccharide-type molecules actually make other tastes more powerful: lemonade with saccharides tastes sweeter and a bit more sour. Add a saccharide to denatonium, and it tastes much, much more bitter.

Your tongue has different groups of taste buds that taste things that are bitter, sweet, sour, and salty. All the foods on this page are either bitter, sweet, sour, or salty. Try to figure out where you'd taste each one on your tongue.

If there's a little saccharide in your soda pop, you'll probably ask for seconds. A little denatonium saccharide in your soda pop, and you'll be asking to be excused!

156

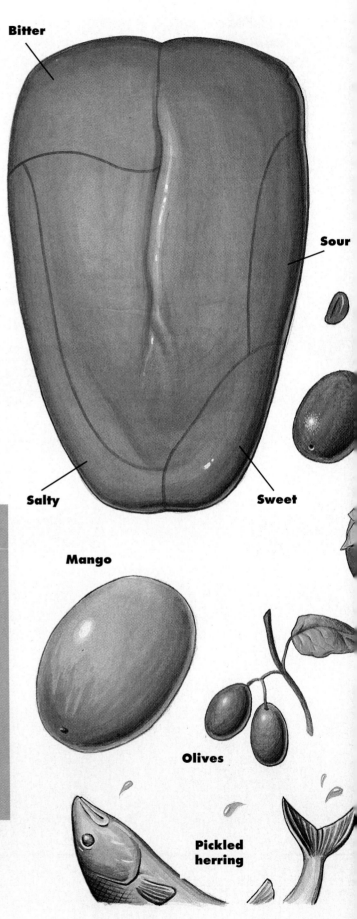

Bitter

Sour

Salty

Sweet

Mango

Olives

Pickled herring

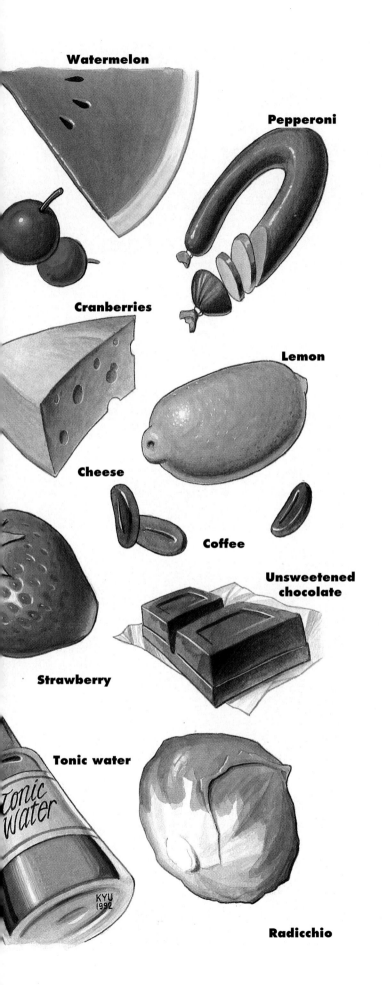

Watermelon

Pepperoni

Cranberries

Lemon

Cheese

Coffee

Strawberry

Unsweetened chocolate

Tonic water

Tonic Water

KYU 1992

Radicchio

EAT UP!

Can you imagine a day when you eat a sandwich down to the last crumb, including the packaging that was keeping it fresh?

Well, that day may soon arrive. A flavorless, edible coating, sprayed on the filling to keep the bread from getting soggy, is not far off, says food scientist Attila E. Pavlath. There is already a spray-on, edible film that can keep a cut apple fresh for about five days.

Naturally, there are still some problems. Pavlath hasn't figured out a way for the film to be applied. So far he has to hand-dip each piece of food himself — and it's hard to get a thin, even coat. The big problem right now is perfecting a film that will keep the moisture out.

But one day soon, you may be reaching for, and chewing on, the edible-film-covered fruits of this scientist's labors!

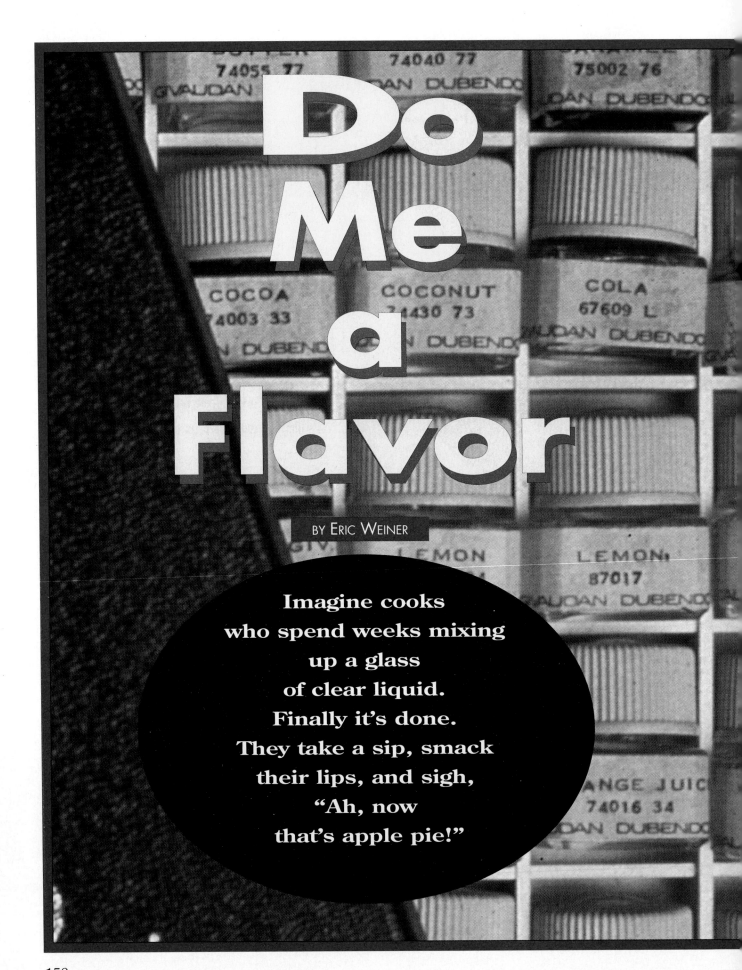

Do Me a Flavor

BY ERIC WEINER

Imagine cooks
who spend weeks mixing
up a glass
of clear liquid.
Finally it's done.
They take a sip, smack
their lips, and sigh,
"Ah, now
that's apple pie!"

They're not kidding. The liquid does taste just like apple pie. That's because these are no ordinary cooks. They're flavor chemists — scientists who use a lab as a kitchen to cook up new tastes.

Artificial flavors — flavors made by a person instead of by nature — cost much less than natural flavors. Companies that make foods use artificial flavors to spice up almost every food product sold in the supermarket.

Copying nature In a flavor chemist's lab, thousands of tiny amber-colored bottles line the walls. Each bottle contains a chemical with a different smell. Why? Because most of what we call taste is really smell.

That's easy to prove. Pinch your nose tightly. Then bite into a piece of pizza (or a peach, or any other food). Since you can't smell it, the food will seem to have little taste. In fact, the tastebuds in your tongue can tell you only whether food is salty, sweet, sour, or bitter.

A single taste may be made up of over a thousand different smells! So how do flavor chemists know what chemicals to mix? Thanks to new and better equipment, chemists have been able to identify more and more of the chemical ingredients in different foods. By trial and error, flavor chemists try to figure out which of these chemicals make up the food's taste.

Clearly, a knowledge of chemistry is a must. "If you're trying to make apple flavor," says chemist Steve Lovis, "you have to know how much of each chemical is in a real apple. If you put in too much of one chemical, your apple could turn into a cherry!"

Who nose best?

The flavor-maker's work is done. He or she has just concocted a new brownie taste. The flavor is mixed into brownie batter and baked. Who takes the first bite?

Most likely it's a taste-testing expert. Some have been sniffing and tasting new products for over 30 years. While most eaters just say "Yum!" or "Yuck!" experts can describe a smell or taste in great detail. Here's one taste-tester beginning to describe a taste of orange sherbet "Dry, tongue sting, bitter, sour, soapy, almost rubbery, vanilla, terpy …" (Terpy means the sting you get when you bite into an orange peel.)

How do tasters develop such a special sense of taste? They practice with other taste-testers for hours. One person might take out three unlabeled bottles of smells. The other one has to name each odor. And there are thousands of them.

Learning to be a good taste-tester takes lots of practice. But there are some tricks. When

Research flavorists Carmelita Ventura and Charles Weiner will keep testing to make sure their candies have just the right fruit flavors.

tasting a lemony soda pop, for instance, testers first sip lemon juice. The intense lemon juice briefly tires out the tester's ability to taste lemon. So when testers taste the soda pop, they don't taste lemon. That way they can taste the background flavors.

Try being a taste-tester yourself. First sip some lemon juice. Then sip some lemony soda pop. What do you taste?

Snack time

You're sitting in a tiny, windowless room. The air around you is odorless. Suddenly, a small sliding door opens and a tray is pushed in. On the tray are a brownie, a pencil, and a piece of paper with a lot of questions.

Your job: Taste the brownie. Then answer questions such as, "Did the brownie have too many nuts?"

This may sound like a strange way to eat a snack. But at a taste-testing lab, it's standard practice. Companies hire taste-testing labs to find out what the public thinks of their new recipes.

Janet Curtis helps run a taste-testing lab. She explains why everyone at a taste-testing lab tastes or sniffs alone.

"We don't want you to be able to see other people taking the same test. If you see someone eating a brownie and making a face, you might decide you don't like your brownie either!"

160

because air odors change the way the food tastes. If the air smells like cabbage, the brownies may not taste so good.

What happens if the brownies do taste awful? "You always get a cup to spit the food into," says Curtis. "You don't have to swallow it. But swallowing gives you more of the full flavor, because you smell the food better that way. So if you don't swallow, we ask that you at least slurp." Slurping passes air over the food and carries the smell up to the nose.

The results from testing labs help companies improve their recipes. But according to Curtis, to get accurate results one has to control the environment.

"We use a special filtered-air system," she says, "to prevent air odors." That's important

Of course, when a lot of people spit out the brownie, it may be time to change the recipe. Companies use the labs to find out what works and what doesn't. They're hoping that when the brownie arrives in your kitchen, you'll slurp, swallow, and ask for more.

MICROWAVE FLAVOR

What's that smell in the lab? Mmm. Roast chicken — but not quite. A timer rings and lab technician Nazhat Forage takes a test tube of brown liquid out of a microwave oven. She records the smell and color, and hands the tube to Professor Varoujan Yaylayan. He smells it. Neither of them thinks it's ready yet.

Yaylayan is a chemist at McGill University in Montreal, Quebec. He is working on experiments to make microwave foods taste better. A lot of them just don't taste right, so it's Yaylayan's job to create flavors that do taste right so they can be added to the food.

Professor Varoujan Yaylayan

Food gets its smell, flavor, and color from the combination of its proteins and sugars. In microwaves, the two don't always mix completely, so the food doesn't smell, taste, or look quite right. Yaylayan works with a certain number of sugars and proteins, mixing and matching them until he gets the result he is looking for.

So far, he's on the right track. He and two other professors, Jim Smith and Hosahalli Ramaswamy, have already come up with a chemical to add to microwave bread dough so that the bread smells like it's being baked fresh in an oven!

— Janice Paskey

CHECK IT OUT!

Does your food taste the same when you have a cold? Ask other people if they have noticed a difference when they have a cold. Why do you think there's a difference?

162

Dessert in a Can

BY THERESA ANN STEVENS

You don't have to be a chemist or even know how to cook to play with chemistry in the kitchen. Ice cream is easy to make. It's pure fun and pure science — and you can eat it!

You will need

- *two coffee cans — about 400 g and about 1 kg — with tight-fitting lids*
- *two litres of crushed ice or small cubes of ice*
- *125 mL of rock salt*
- *500 mL of half-and-half cream*
- *7.5 mL vanilla*
- *125 mL sugar*

The result will be 500 mL of smooth and creamy ice cream.

Mix the cream, vanilla, and sugar together. For an extra-special flavor, try adding 75 mL of fruit, 3 of your favorite cookies (crushed into small pieces), or 50 mL of chopped nuts.

Pour the mixture into the smaller coffee can. (The can will be about three-quarters full.) Place the lid on tightly and put the small can inside the larger can.

Starting with ice, alternate layers of ice and salt between the outside of the small can and the inside of the large can, packing the ice down as you go.

When the layers get to the top of the small can, cover the entire lid with ice. Then place the lid on the large can, making sure it fits tightly.

Set the can on the floor and gently roll it back and forth. You may need to wrap a small towel around the outside of the large can so it will not stick to your hands. Roll the can for at least 10 minutes. The ice inside the can will get watery, but keep on rolling!

After rolling, put the can upright. Remove the towel and lid from the big can, and slowly pull the small can out of the big can. Wipe the ice and salt from the lid; then remove the lid carefully.

The ice cream will be frozen to the sides and bottom of the can. (If it's not thick enough, put the can in a freezer. Check it every 10 minutes or so.) When it is thick enough, stir the icy parts off the sides and bottom into the mixture until it's smooth and creamy.

So how does this all work? To freeze the mixture, you need water that is colder than 0°C. But that's the temperature of water when it becomes ice. The trick is the salt. It melts the ice into water and cools the water to about -3°C. The below-freezing water cools and freezes your ice-cream mixture!

Did you also notice that you had a fuller can when you were finished? Rolling the cans adds air to the mixture, making it take up more space. The water in your mixture takes up more space as it freezes, too.

You've just finished experimenting with science. And even better — you get to eat your experiment, too!

CHECK IT OUT!

There's science in other treats, too. Popcorn kernels, for instance, have water inside them. That's what makes them pop when they're heated. Find out more about this neat treat.

FLASH**BACK**

Food from Afar

In the early 1900s you would have had to live on a farm to get fresh eggs, or by the sea to have fresh fish. Meat, fish, fruits, and vegetables couldn't be shipped long distances or kept sitting around for days. They would spoil unless they were pickled, salted, dried, or preserved. But thanks to Mary Pennington, you can now eat fresh food, in any season, that comes from thousands of kilometres away. Pennington was a bacteriologist who was fascinated with the problem of refrigeration. She began her career studying milk. She did the first studies ever on preserving perishable foods, or foods that could go bad, in cold storage. Pennington went on to develop national standards for milk inspection.

She became head of the United States Department of Agriculture's new Food Research Laboratory, and was soon setting standards for everything from the insulation in railroad ice cars to the type of knife that allowed clean butchering of meat.

Pennington wasn't the type to sit still. She became well-known to shipping and packaging officials. She also spent lots of time riding on the trains to study railroad refrigeration first-hand — and it paid off. She and her staff developed and improved food-processing methods, as well as shipping and storage methods. Her techniques made food spoil less and taste better.

So the next time you sink your teeth into a juicy apple in February, think about the long trip the food took to get to you — and the woman who invented a way to keep it cool and fresh on its way!

Get Cracking!

Fry 'em, poach 'em,
beat 'em, boil 'em —
they're still eggs,
no matter what you do
to them and no matter how
different they look.

But why does
a scrambled egg look
so unlike a boiled egg?

Eggs are mostly water but they're also full of protein. That's why they're good for you. That protein is packed into the egg in long, folded-up chains, much too small for you to see. But when an egg is heated, those chains, or molecules, of protein start to move quickly and unfold. If they bump into other protein molecules, they get attached again, but not the same way they were before. As the egg continues to cook, the proteins get more and more bonded together until they can hardly budge — they're stuck solid. You can't watch the molecules move but you can see the liquid yolk and white turn solid. There! You have a well-done fried egg.

BY ELIZABETH MACLEOD

166

Fluffy foam

Meringues (MER-angz), *soufflés* (soo-FLAYZ), and angel food cakes all depend on the amazing ability of whipped egg whites to become eight times fluffier than normal. Whipping forces lots of air bubbles into the white. The proteins in the whites unfold and re-attach to make the strong walls of the bubbles.

Beaten egg whites in cakes or *soufflés* rise when they're baked. That's because gases inside the bubbles take up more room as they're heated. But the gases won't pop the bubbles and make your cake collapse. Heating makes the proteins of the bubble walls even stronger — too strong to burst.

One thing that will make the bubbles collapse, though, is even the tiniest bit of egg yolk. Yolks don't foam very well because they are made of a different kind of protein. They also have a kind of fat which stops the bubbles from forming.

All scrambled

Creamy scrambled eggs are eggs with really mixed-up whites and yolks. The white contains much more water and less protein than the yolk, so it starts to cook at a lower temperature than the yolk.

When eggs are well blended, the whites and yolks are so broken down that you don't notice the difference in how they've cooked. But when the eggs haven't been mixed enough, the pockets of white and pockets of yolk cook separately into lumps.

168

Whip up an edible experiment in your kitchen. Stir up a mixture of three eggs, 30 mL milk, and some cheddar cheese. With an adult to help you, place this mixture into a hot frying pan over medium heat and watch the bottom change as it cooks.

Experiment by throwing in some seasonings or chopped vegetables (tomatoes or mushrooms are good) and flip half of the omelet over so that it looks like a semi-circle. Let it cook for a few more minutes and you are ready to serve your masterpiece.

How can you tell the difference between hard-boiled and raw eggs without cracking them open? Raw eggs float in water and hard-boiled eggs sink!

Ring around the yolk

The inside of an egg is white, yellow — and black?! The black ring you sometimes see in hard-boiled eggs is caused by a combination of iron and sulfur. The iron comes from the yolk and the sulfur comes from the white. As the egg gets heated, the sulfur is released. When it comes in contact with the yolk, it mixes with the iron to make a chemical called ferrous sulfide, which just happens to be dark. If eggs are boiled only as long as needed and cooled quickly, the chemical doesn't have a chance to form.

Dry up

Spoonful of dried eggs, anyone? Who uses dried eggs? They're being used more and more in the baking industry since they are so easy to store. But preparing dried eggs is a long process.

First the eggs are broken into a cool vat, then they are pumped to a dryer. There, spray nozzles turn the liquid egg into a fine spray of tiny particles that are forced into a very hot chamber to be dried. The powdered eggs collect at the bottom of the chamber and then are cooled by passing through a system of coils. Then they're ready to be packed into barrels. They can be stored for a long time, as long as they're not exposed to air.

Eggs can also be freeze-dried — that's how astronauts eat them.

EGGS IN SPACE

Hot scrambled eggs for breakfast? No problem — unless you're an astronaut in space. In that case, you have to poke a special needle into your pouch of freeze-dried scrambled egg powder, squirt water in through the needle, squish the pouch with your hands to mix the water and eggs, then warm the pouch in a special oven.

To eat, you'll need scissors to cut open the pouch and a spoon, because food sticks better to it than to a fork. Squeeze on some liquid salt (regular salt would float in space) and dig in. Mmm!

The first space food came in tubes that astronauts squeezed directly into their mouths. It looked and tasted pretty bad. The foods are better now and some are the same as you eat: rye bread, cookies (tiny so they can be eaten in one bite and leave no crumbs), and peanuts.

Some snacks that are easy to prepare on Earth pose special problems in space. How many hands do you think you'd need to prepare a peanut butter sandwich in space? Astronaut Sally Ride says three hands are just right!

Want to try some space-type food? Try the lasagna or chili sold off the supermarket shelf (no refrigeration necessary!) in flat pouches.

Astronaut Rhea Seddon digs into some space food!

Super

BY DON HERBERT

**Walk down the baking aisle of your supermarket.
See any pure chemicals? You wouldn't think so —
who would want to bake with chemicals?
Well you can and, in fact, you usually do.
One of them is baking soda. It's one of the few
chemically pure substances in the
supermarket. And baking
isn't all it can do.**

Baking soda, or sodium bicarbonate (by-CAR-buhn-ate), is an important ingredient in making cookies, breads, and cakes. Baking soda dissolved in water reacts with acids and makes carbon dioxide bubbles. Those bubbles make your baking rise. Bakers use acids such as vinegar and sour milk to react with the baking soda. That way cakes, cookies, and breads turn out light and fluffy.

Baking soda is also one of the main ingredients in baking powders, so not surprisingly, they act much the same way. The gas bubbles they form help the baked goods rise.

Soda

When it is dissolved in water, baking soda is a base. That's the opposite of an acid. You can dissolve a spoonful of baking soda in water and drink it to calm indigestion, which is caused by too much acid in your stomach.

Baking soda is made of very fine particles. In the form of a paste the particles act as a mild scrubber — for cleaning everything from teeth to countertops. If you get the sticky stuff from an evergreen tree, called resin (REZ-in), on your hands, wet your hands, rub the spot of resin with baking soda, and then rinse. The resin will be gone.

In dry form, baking soda absorbs moisture as well as chemical vapors (smells!) that may be in the air. That's why it works as a deodorizer in the refrigerator.

Baking soda doesn't burn. Heating it lets off carbon dioxide, which doesn't burn either. So baking soda is sprinkled on grease and oil fires to put them out. It's also one of the main ingredients in soda-acid fire extinguishers.

Baking soda may look ordinary, but now you know — it's really extraordinary!

Soap's Up!

Ready to play a round of golf? You won't get too far with these balls — they're really soaps! All of these unusual soaps are made by a company in Vermont.

All aboard the tub train! These soaps are a tubful of laughs!

The company's best-selling soap is Garfield, the lasagna-loving comic-strip cat. Teddy bears are popular, too. These soap shapes look so real that you might pick up a soap telephone to make a call!

Before work begins, the company buys soap base and perfumes. Perfume formulas are a closely guarded secret, but the recipe for soap is well-known. For centuries, people have mixed fats or oils with a strong chemical called lye to make soap. The soaps you see here have coconut oil, lye, and an animal fat called tallow as their base.

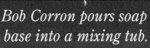

Bob Corron pours soap base into a mixing tub.

174

Coloring is ladled out to be added to the soap mixture. Several shades might be combined to get the final one. A mixer will blend all the ingredients together.

It looks like a modern sculpture, but this is a refiner. It does further mixing. As the ingredients warm up in the refiner, they stick together to make soap chips.

Molded soap crayons travel by conveyor belt. Joan Bricker trims them as they come off the line.

175

MIXED-UP MIXTURES

Scientists say that something that brings together two things that like to stay apart is called an emulsifier. Soap is an emulsifier (ee-MUHL-si-fy-er). It brings together water and oil — in your laundry, in your sink, on your face — to wash away dirt. But did you know you can eat an emulsion (that's what the liquid is called when it's all mixed together), too?

Salad dressing is an emulsion. To make salad dressing you mix oil with vinegar, but you have to add something else, because the two don't mix. Some popular additions to salad dressings are eggs and mustard. Actually, salad dressing is only a temporary emulsion. You have to add plant matter, such as lettuce, to make it stay stuck together. Otherwise, it separates and you have to shake it again to mix it up.

Make your own yummy mixed-up mixtures. Experiment with ketchup, mayonnaise, pickles, soya sauce, or anything else you think would be good to make your own secret sauce for burgers.

These layers of oil and vinegar like to stay apart. But there's an emulsifier in the bottle, too. If someone shakes up the liquid, the oil and vinegar will stick together for a minute or two!

Use oil, vinegar, mayonnaise, or seasonings to make a super salad dressing. Make up some new mixtures for barbecue sauces or sandwich spreads. Have science for lunch!

THE DeeP
Blue Sea

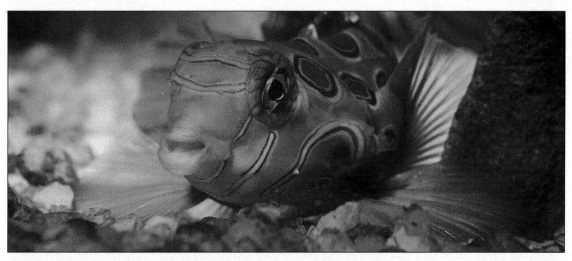

Bull Kelp

The Perfect Fishing Line

People who fish usually use lines made of nylon or steel. But if they really want a fabulous fishing line, they should be using seaweed.

In the late 1700s European explorers in North America reported that the Kwakiutl (kwa-kee-OO-tuhl) who lived on the west coast of Canada used seaweed for fishing line. They used bull kelp, which is a brownish plant that attaches to the seafloor and stretches up to the surface of water. Although it grows close to shore, the stems can sometimes be 10 m long. The leaflike blades, called fronds, float on the water.

The explorers, and others who followed them, were intrigued that the

BY JAY INGRAM

Kwakiutl didn't just twist the seaweed and use it as instant fishing line. Far from it. They first laid the stems in fresh water and anchored them with stones. After a few days the kelp had turned from shiny brown to white, and it had swelled to twice its usual diameter.

The Kwakiutl then stretched the kelp as tightly as possible on a wooden frame for 10 to 12 days. During that time the stalks shrank to about three millimetres in diameter. At this point the kelp was ready to be used, although sometimes it was hung over a fire first to blacken it, so the fish wouldn't be able to see it. Just before they actually used it, the Kwakiutl would soak the line in salt water to make it bend easily.

And it worked. In fact, the Europeans who first reported it admitted that it worked a lot better than the fishing lines they used. The Kwakiutl line wouldn't twist or get kinks in it, and if it was thrown overboard, it wouldn't get tangled. But no one bothered to ask why the Kwakiutl went through this elaborate preparation. You can bet that no one thought it had anything to do with chemistry. That's right — the Kwakiutl were using chemistry to make the sea-weed into a better fishing line.

A scientist in Chicago named Michael LaBarbera decided a few years ago to find out what happens to stalks of bull kelp when they are soaked in fresh water, stretched, and dried. He found that putting the freshly picked kelp stems into fresh water by itself does nothing. But soaking in fresh water followed by drying — as the Kwakiutl did — changes the seaweed dramatically. It becomes twice as

strong as it was before. But it's even better to stretch the stalks while they are drying. They are then five times stronger than they are when just freshly pulled from the ocean. Bull kelp stalks are just about as good as modern fishing lines, which of course are made from synthetic materials such as nylon.

LaBarbera thinks that soaking the stalks in fresh water and then stretching them causes chemical changes that make the seaweed stronger. Stretching also pulls on fibers inside the seaweed, straightening them out. That helps the stalk resist breaking when there's a fish tugging on it.

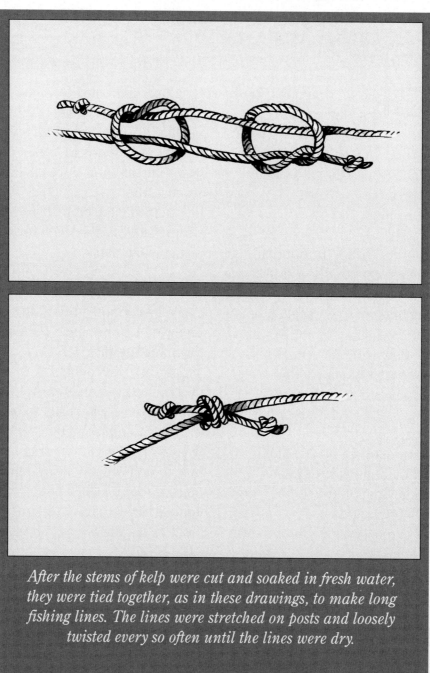

After the stems of kelp were cut and soaked in fresh water, they were tied together, as in these drawings, to make long fishing lines. The lines were stretched on posts and loosely twisted every so often until the lines were dry.

We always think of today's inventions as the most advanced. But hundreds of years ago, the Kwakiutl were making a strong, recyclable, biodegradable fishing line, entirely from natural materials.

LIFE

ould you like to live near a 400°C hot spring that stinks like rotten eggs? Not likely. But ocean scientists have recently discovered species of shellfish, worms, sponges, and other animals who make their deep-sea homes in such cozy spots. They live at hot-water vents.

Dr. Verena Tunnicliffe dives into the dark, mysterious world of those vents in a tiny deep-sea submersible. She's exploring parts of the ocean floor off the west coast of North America. Using the submersible's robotic arms, Tunnicliffe collects samples of the strange creatures living near these hot springs.

Most animals get their energy from the sun. But the sun's light doesn't get to the bottom of the ocean. Down there, it's very dark

Dr. Verena Tunnicliffe

and, in most places, very cold. In some places, though, the Earth's crust has shifted, leaving narrow cracks on the ocean floor. Seawater seeps into the rock around these cracks. When it gets to the intense heat of the Earth's interior, it shoots out of the cracks, forming deep-sea hot springs.

Hydrogen sulfide (that's what gives off the rotten-egg smell) is shot out with the seawater. It's usually poisonous, but hot-vent animals depend on hydrogen sulfide to live.

Tunnicliffe and other scientists carefully study the samples she finds, searching for clues to the amazing story of their survival. I talked to her about her fascinating work.

ON THE HOT *Spot*

BY CYNTHIA PRATT NICOLSON

181

Cynthia: When did you first decide to become a marine biologist?

Verena: When I was young I was intrigued by the seashells that my mother collected. In high school I always chose projects about ocean life and my interest got stronger and stronger. But I grew up in Ottawa, Ontario, so I never saw an ocean until I was 19!

Cynthia: What made you pick hot-water vents for your research?

Verena: It happened by chance. I had studied mud flats on the east coast and coral reefs in Jamaica. In 1982, while studying small undersea mountains, my ship got lost off the west coast. Luckily our boat met a research ship from the University of Washington. The scientists on board were dredging the ocean floor, searching for hot-water vents. I just happened to be there when they pulled up samples of sulfide compounds and tube worms — two signs of vent activity. The only other hot vents that had been discovered up until then were much farther south, off the Galapagos Islands and the coast of Mexico. This was a totally new system.

Cynthia: What happened after that first discovery?

Verena: Because I had lots of experience with deep-sea diving in submersibles, I was asked to lead a joint Canadian-American expedition to explore the hot-water vents off the west coast. We started out with a simple question — would we find the same kind of animals here as had been found at other hot-water vents?

A bright red octopus and a rat-tail fish (about the length of your arm) feed near a hot-water vent on the ocean floor. If you check out the bits of empty clamshells scattered around, you'll know what the octopus had for dinner!

182

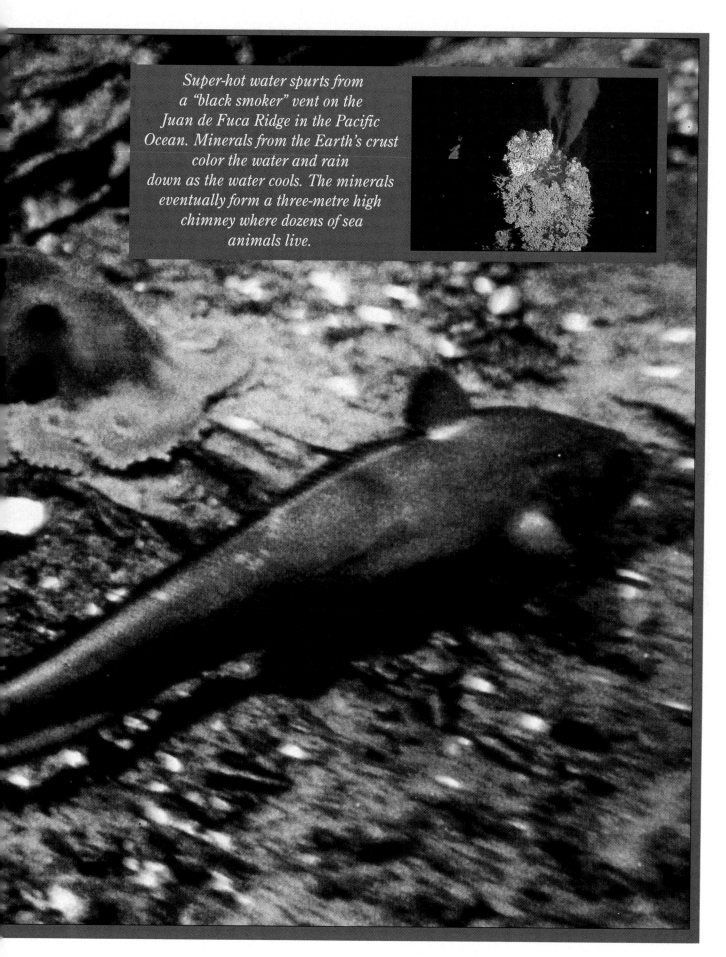

Super-hot water spurts from a "black smoker" vent on the Juan de Fuca Ridge in the Pacific Ocean. Minerals from the Earth's crust color the water and rain down as the water cools. The minerals eventually form a three-metre high chimney where dozens of sea animals live.

Cynthia: Why did you need to use submersibles?

Verena: Because the vents are 2000 or 3000 m below the surface and the water pressure is enormous. A scuba diver would be crushed to death long before reaching that depth.

Cynthia: What kinds of creatures did you find?

Verena: Fish, crabs, giant clams, octopuses, snails, tube worms — I've got a list of over 300 species, most of them brand new to science.

Cynthia: Hot-water vents have now been found all over the world. Do the same kinds of animals live near all of them?

Verena: There are some differences but generally the animals at these vents are more like one another than they are like animals in the rest of the deep sea. For instance, the crabs at hot-water vents all over the world are still very much like their early ancestors. Crabs in other parts of the sea have changed by leaps and bounds.

Cynthia: Why is that?

Verena: I believe it's because vent animals have been separated from other living things for a long time. While the rest of the world experienced major changes — asteroids hitting Earth, dust clouds, tidal waves, and so on — animals at the hot vents have carried on in their own little systems. The ocean has protected them from all the action up above. These are the only ecosystems on Earth where things have remained much as they were 80 to 100 million years ago!

Hot-vent tube worms, which sometimes grow to 1.5 m, attach themselves to a seafloor ridge off the coast of Mexico. Nearby, small crabs and fish dart about. Like the tube worms, these species are found only at hot-water vents.

The next time you're at the ocean — or a lake, river, or pond — do some investigating of your own. Choose a spot and take a look at the plants and animals that live there.

Why do you think they live there? How do you think they might have adapted to this habitat? Keep a journal of your observations.

Scuba is a short-form for self-contained underwater breathing apparatus.

FANTASTIC FACES

If you were to design a really weird-looking fish face, it might not be any more weird than the faces of a lot of living sea creatures. Many fish have features that seem odd to humans. Yet these features often serve useful purposes. You can learn interesting things about fish by studying their faces.

A face may give clues to where a fish lives. Strange shapes or colors may help the fish blend in with its surroundings. This camouflage gives a fish protection. Enemies can mistake the camouflaged fish for leaves, rocks, seaweed — or even for other kinds of animals. You can often guess where a fish lives by the things it seems to match. For example, a sand-colored fish probably spends its time on the sandy seafloor.

A brightly colored face may serve a different purpose. It may help the fish advertise itself. A recognizable color pattern makes it easy for a fish to find others of the same kind. Bright colors on the face may also help protect a fish. Some colorful fish taste bad. A few can even be poisonous. Their colors may warn other fish to stay away.

The goosefish is also called the allmouth — its mouth is often as large as a dinner plate. The goosefish eats almost anything, including small sharks, squid, crabs, other fish, and even diving birds. It grabs passing meals with needle-sharp teeth. The growths that look like weeds under the mouth are bits of skin. They help camouflage the fish when it rests among seaweed on the ocean floor.

186

Tool or weapon? The sawfish's snout is both! Edged with toothlike outgrowths of the fish's scaly skin, this long, flat blade is strong and dangerous. Sometimes the sawfish uses it to dig into the sandy ocean floor, stirring up a meal of worms and snails. Other times, the hungry sawfish flails at a passing school of fish, stunning and killing its prey. Though it is as long as a full-sized racing car, the sawfish sometimes needs to defend itself, too. When a hammerhead or a great white shark threatens, the sawfish slashes and jabs with its snout. With enemies like these, it helps to have a face like a chainsaw blade!

You might say that the psychedelic (sy-kuh-DEL-ik) fish has good looks and bad taste. The brightly colored creature belongs to a group of fish called dragonets. The group includes some of the most brilliantly colored fish in the world. Mucus that covers the body of the psychedelic may taste bad to other fish. Some experts believe the psychedelic's bright colors warn other creatures not to take a taste.

Try THIS

Design and draw your own wild fish for a specific environment. How does it eat? How does it protect itself?

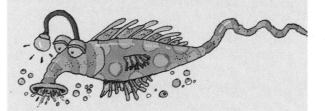

CHECK IT OUT!

Some fish are fantastic feeders, too. The starfish throws out its stomach to get food, and the barnacle throws out its feet! Visit a pet store and talk to someone who works there about the kind of fish the store has and how those fish eat. Or visit a zoo or an aquarium and ask the keepers there how their fish eat.

The sea urchin walks on the tips of its teeth.

189

Where's the deer-crossing sign?
What look like antlers on the head
of this small fish are really growths of flesh
called cirri (SEAR-eye). The cirri help
the fish blend with the surrounding sea
plants, corals, and jagged rocks.
This fish is a yellowfin fringehead.
It belongs to a group of fish called
blennies. They spend most of their time
hiding in shells, in holes dug by
other animals, or in old bottles.
Most leave their hiding places only
to feed on tiny plants and animals,
to mate, and to lay eggs.

FLOATING FOSTER MOM

Sea otter Pico swims with Julie Hymer, his foster mom.

At the Monterey Bay Aquarium, a wet-suited foster mom swims with a sea otter pup called Pico.

Pico, an orphan, was rescued by the aquarium when he washed ashore. His human foster mom taught Pico everything his natural mother would have taught him, from diving to finding and collecting food. She even showed him how to sleep wrapped up in kelp strands in the water.

At last, when the foster mom thought Pico was ready, she took him to visit the kelp beds where the wild sea otters live. The more he visited the kelp beds, the more reluctant he was to return to the aquarium. Finally, Pico was ready — he left his foster mom and started his new life as a wild sea otter.

— Sylvia Funston

Under Antarctica

BY KATHY ANN MILLER

Have you ever dreamed of exploring another planet? I did, and my dream came true in Antarctica, the frozen continent at the South Pole. I spent three summers in Antarctica at McMurdo Sound, south of New Zealand, exploring another planet without ever leaving ours.

Miller holds up the wetsuit that's going to keep her warm under the ice.

I'm a marine biologist in Puget Sound, Washington. Seaweeds are my speciality. I learned to scuba dive to see where the seaweeds live and to study how they fit with other creatures in the undersea world.

It's November, early summer in the southern hemisphere. Today is clear, cold, and sunny. My diving partner, Mary, and I load our gear on to a truck that has tracks instead of wheels, and set out. The sea ice in Antarctica is between two and seven metres thick, so it makes a good, solid road. It takes about two hours to reach our site. Where's the water? Under our feet!

You can see the sight of our dive — Cape Evans — on this map. On a globe, this area would be right at the bottom!

We have to make a hole in the ice. This morning we use dynamite and the blast is really loud. Sometimes we drill through the ice instead. Next, we wedge a bamboo stake in the ice at the edge of the hole, tie on a thick rope, and drop it into the water. The rope is weighted to help it fall, and it's marked at each metre. We watch the marks fly by until the rope hits bottom with a thunk. Now we know how deep the water is.

193

To keep warm and dry in the cold water, I wear long underwear, a thick quilted insulating suit, and a rubber-coated canvas dry suit. The dry suit has rubber seals at the wrist and neck that keep water out, built-in boots, and a long zipper over the shoulders. I buckle on a weight belt, too. Once I'm under water, I'll adjust the amount of air in my suit so that I'll be weight-less, but first it will take almost 20 kg of lead to sink all this insulation!

when suddenly I'm out of the ice tunnel! I'm below the hole and the world has opened up under the ice. I look up to see Mary following me down, fins emerging from the hole, and then … she's out, too.

I'm at the edge of the hole. My helper has strapped my double air cylinders to my back and connected the hose to my suit. A gauge on my wrist will tell me how deep I am. A pressure gauge on a hose attached to my air cylinders will tell me how much air I have left. I'm wearing gloves that strap and seal over my wrists and keep my fingers warm and dry. In my 55 kg of protective gear, I look like a fat seal!

All dressed, we slip into the narrow, two-metre-deep hole one by one.

Yow! I never get used to that first blast of ice-cold water on my face. I think warm tropical thoughts as I glide past the white sides of the hole. I look at the layers of crystals in the ice, check my watch and gauges, feel my face going gradually numb …

The ceiling of sea ice looks like a vast fluorescent light — and the brightest light is the diving hole.

I hear the fizz of bubbles and the echo of my breath burbling through my regulator. Before me is an endless landscape of gently rolling sea floor, large boulders, a long valley falling down as far as I can see. Above me is a huge, blue ceiling — the sea ice.

I can see a straight white line, the road that we drove on to reach here. I can see the bright light of the diving hole. Far in the distance, I see a long crack where the ice meets the shore. I stop thinking about the cold. I'm floating in space!

As I fall, I add spurts of air from my hose to my suit. That keeps me moving down at an even rate, and adds insulation to my quilted underwear. Without added air, the pressure would squeeze my suit tight. Mary and I signal "OK" and then clap our hands, our private signal for "Here we are! Can you believe it?"

As I swim over the bottom, the water feels almost thick, like honey. I approach a huge pile of boulders and take a closer look with my flashlight. The rocks are covered with animals without backbones, or invertebrates. I see white and yellow sponges, apricot soft corals, delicately branched byrozoans, and lovely frilled sea slugs.

At the base of the boulders, on the gravel bottom, is a mosaic of small red and orange sea stars, with purplish spiny sea urchins here and there. Big white snails. Metre-long smooth worms in coils. Big pale sea anemones (uh-NEM-uh-neez). And, on smaller rocks, tufts of deep red seaweed!

The creatures who live here all have one thing in common — they move very slowly. Their bodies have adjusted to life at very low temperatures. Even the fish lie on the bottom and won't move until you poke them!

195

These creatures have no backbone.
The one that looks like a feather
is called a feather star.

196

Today we're collecting samples of seaweed and of sea urchins. We want to see if the urchins eat the seaweed, as they do in the oceans back home. We compare depth gauges: we're 15 m below the surface. (The bottom here is not as deep as in some oceans. We're on the continental shelf that surrounds nearby Ross Island.) Mary holds the bag as I chisel clumps of weed off the rock and drop them in. I count out urchins and drop them in, too. When we have enough, she carefully closes the bag, clips it to my belt, and we check the time. We have seven more minutes of the 20 we planned.

Our ears are filled with an unearthly music of long peals, warbles, and clicks. We stop and look around. In the direction of the tide crack along the shore, we see the huge shadowy forms of Weddell seals. In a flash they're here, swimming and circling gracefully. We are surrounded by their calls and their acrobatics. One female comes close enough for me to see her big eyes roll and check us out. As suddenly as they appeared, the seals are gone. We look at each other and clap our hands: "Here we are! Can you believe it?"

We look at our watches and swim to the rope below the hole. We slowly go up, letting air out of our suits to control our ascent. Here's the hole. I'm back in the smooth white tunnel. My head pops out and there's my helper, kneeling at the edge, waiting to help me out.

Wrapped in my warm parka with a hat on my wet hair, drinking cocoa, I put our finds in an insulated bucket. As we get ready to drive

ice fish floats by. back to the base, I can hardly believe I'm back on Earth.

That cold watery world with seal calls, amazing plants and animals, that glowing blue ceiling — was it all a dream? Yes — a dream come true!

Try THIS

Which freezes first, fresh water or salt water? To find out, get two containers and fill one with fresh water and the other one with salt water (stir in about 50 mL of salt until it's dissolved).

Place both containers in the freezer and check them every half hour or so to see which freezes first.

Each day 7 500 000 t of water evaporate from the Dead Sea in Israel.

197

FLASHBACK

Sharing the Ocean

Like most people, you've probably never been to the bottom of the ocean. But you probably know a lot about it, thanks to the work of Jacques-Yves Cousteau. Cousteau not only brought the amazingly beautiful world of the deep into everyday life, he also created much of the equipment that made that possible. In 1943 he helped invent the Cousteau-Gagnan Aqualung, a device that lets divers breathe underwater. The idea came from a car engine valve that he adapted!

The Aqualung is a canister of compressed air strapped to the diver's back, with a tube leading to the diver's mouth. It pumps air into the diver's lungs so that the diver can keep breathing just as though he or she were on the surface. Until the invention of the Aqualung, divers had to use bulky breathing equipment that was difficult to move around in. The Aqualung made it possible to observe coral reefs and ocean life much more easily. Today scuba divers use an Aqualung that hasn't changed much since it was invented.

But Cousteau didn't want to stop with short dives. He wanted scientists to be able to study the sea over longer periods of time. He developed an underwater vehicle called the *Soucoupe*. It could dive to 185 m and stay under for 20 hours. He also developed the first underwater diving station. There scientists could live for several weeks, more than 12 m below the surface, while they studied the life of the continental shelf. The shallow waters around the shelf are perfect for life forms because lots of sunlight gets in. Cousteau had hoped that someday his diving station would be helpful in setting up human colonies along the continental shelves!

Cousteau was born in France in 1910. Since 1951 he has explored the world's oceans aboard his research ship *Calypso*, sharing the incredible undersea world with people everywhere through his books, films, and television programs. He has fought hard to stop France from dumping radioactive waste into the Mediterranean Sea, and he continues to work to help the oceans stay beautiful and healthy.

WAVES
Wild

What did the ocean say to the shore? Nothing, it just waved — as it's doing here for professional surfer Mark Foo. This surfing hot spot on the North Shore of Oahu in Hawaii is called "Off the Wall." Surfers come from all over the world for the powerful waves. On popular surfing beaches like this one, waves are often one and a half to three metres high. But sometimes the ocean delivers really big ones, up to 10.5 m.

For large waves to form, the wind far offshore must blow very hard in the same direction over a large area. It has to keep blowing for many hours or days.

The largest wind wave ever measured at sea occurred in 1933. The crew of a tanker spotted the enormous wave in the Pacific Ocean and took accurate measurements of it. The wave rose almost 34 m — as tall as an 11-story building.

Some scientists think this gigantic wave was a rogue (rohg) wave. The huge monsters known as rogues are the wildest of all waves. While the heights and shapes of most waves can be predicted, rogues take people by surprise. They may form by chance when several waves combine, or when waves meet an opposing ocean current.

Open ocean waves change as they enter shallower water near the shore. They slow down and grow steeper and higher. Soon the tops of some of the waves begin to curl over. And — before the waves finally break into foam — some lucky surfer may catch one and have a wild ride.

Breaking waves come in three basic forms.

1 *Spilling breakers, the most common, appear above slightly sloping ocean bottoms. They crumble gently, bubbling near the tops, or crests.*

2 *Plunging breakers, the most spectacular, occur above slightly steeper bottoms. After curling over, they collapse with loud roars and bursts of spray and foam.*

3 *Surging breakers don't really break at all. They meet a steep bottom just as they are beginning to peak. The bottom reflects some of the waves' energy back out to sea. The waves, which are now smaller, just wash onto the shore.*

How does this surfer catch a ride? As the wave moves, she slips down the wave's front, which is constantly forming. The forces acting upon her are in perfect balance, so she moves forward with the wave at a constant speed. When the surfer tilts the board, she changes the balance of forces, making herself go faster or slower.

Build your own wave-maker. Fill a large tub or baking pan halfway with water. Use different motions to form a variety of ripples or miniature waves in your tank. See if you can find just the right rhythm for a breezy day, a good surfing day, or a scary, stormy day.

The Sargasso Sea has no shore. It is surrounded by the Atlantic Ocean.

OTHER OCEAN ACTION

Playing in the surf isn't all waves and boards — just ask Naomi Ruza. The 11-year-old teaches snorkeling to young swimmers in the Caribbean Sea, off the island of Bonaire.

"Snorkeling is fun and easy," says Naomi. "It's peaceful underwater. The fish, the shells, and the coral are beautiful."

Her students learn to swim underwater breathing through a snorkel, or tube, that sticks up above the surface of the water. They wear clear face masks so they can open their eyes while swimming.

Some of Naomi's students are a bit afraid of the fish they see. They overcome their fears with help from someone their own age. "When they see that I'm not

scared, they feel more comfortable," Naomi says. "There are only two dangers to watch out for near Bonaire: fire coral, a stinging coral, and scorpion fish. And they only hurt if you touch them."

Naomi, in the middle, teaches her students how to use the snorkel gear.

Naomi has been snorkeling since she was five and teaching since she was nine. One day she hopes to become a marine biologist. Before that, however, she plans to achieve another goal. At age 12, as soon as she is old enough to try for junior certification, she wants to learn to scuba dive.

KEEPING CURRENT

Sneakers ahoy! Curtis Ebbesmeyer, an expert on ocean currents, thought something was adrift when he heard that thousands of sneakers were washing up along the shores of the Pacific Northwest.

After doing some research, he discovered that 80 000 sneakers had been washed overboard when a cargo ship got caught in a storm. It was the largest number of floating objects ever dropped in one place in the Pacific Ocean.

Ebbesmeyer thought that the floating footwear could help him with his research. "I've been keeping records of where and when each shoe was found," he said. Ebbesmeyer hopes that each sneaker sighting will put him one step closer to mapping Pacific Ocean currents. In the future, this research may even help scientists predict the path of an oil spill.

Now that's research that's on the right track!

The Rhythm of the Tides

Have you ever explored a wide ocean beach one day, and then gone back another day to find that your beach has shrunk to a narrow strip of land?

Have you crouched down at the ocean's edge to look under rocks for a few minutes and, before you knew it, your feet were getting wet?

If you have, then you have seen something of the tides. Tides are the regular rise and fall of the world's oceans.

BY GAYL HIPPERSON

At low tide, Mont St. Michel, in France, is surrounded by land.

In the middle of the ocean you wouldn't notice the tide at all, but on shore you can see the water climbing up the beach at high tide and dropping back again at low tide. The whole ocean, right to the very bottom, moves with the tides. What powerful force causes this great movement? The tides are caused by our moon and sun.

Just as the Earth's gravity pulls objects to it (which is why things fall down, not up!), the moon and sun pull on the Earth. Their pulling causes the tides. The moon is the main control of the tides. The sun also pulls on the Earth and makes tides, but it is so far away that its pull is not as strong as the moon's.

The moon's pull causes the oceans to bulge outward from the Earth. There is one bulge on the side of the Earth facing the moon and a second on the opposite side of the world. Each water bulge creates a high tide. As the world turns on its axis and the moon moves slowly in its orbit, the high tides circle the Earth daily. The tide rises and falls twice a day on most shores.

The tremendous tugging of the moon moves — believe it or not — the land and the air as well as the ocean. Scientists have shown that the continents are pulled about 25 cm toward the moon, and the atmosphere surrounding the Earth is drawn many kilometres out into space. Even you are pulled by the moon and sun — you gain and lose a few milligrams in weight with the rise and fall of the tide!

The sun also plays a part in the tides. Twice a month, at the full moon and at the new moon, there are extra-high, spring tides. (The name comes from the water springing high onto the shore, not the time of year.) Spring tides occur all year round. When they occur, the Earth, moon and sun are in line with each other and the sun pulls along with the moon.

At high tide, water flows in over this sand and surrounds the building so that it looks like an island.

About a week after the spring tides, when there is a half moon in the sky, the sun and moon are at right angles to each other and the sun pulls against the moon. This tug of war produces smaller than usual tides, called neap tides.

The highest tides in the world are at the Bay of Fundy, between New Brunswick and Nova Scotia. The tide on Nova Scotia's south shore rises only about one or two metres. But at the head of the Bay of Fundy, on the north shore of Nova Scotia, the tide rises an amazing 18 m. Why? Local geography has a lot to do with the way the tides act. The bay is shaped like a funnel. Tides flowing in the deep, wide mouth have no place to go but up as the bay gets narrower and shallower. Also, a natural back and forth movement of the water, like you can make by gently rocking a pan of water, sloshes water higher into the ends of the bay.

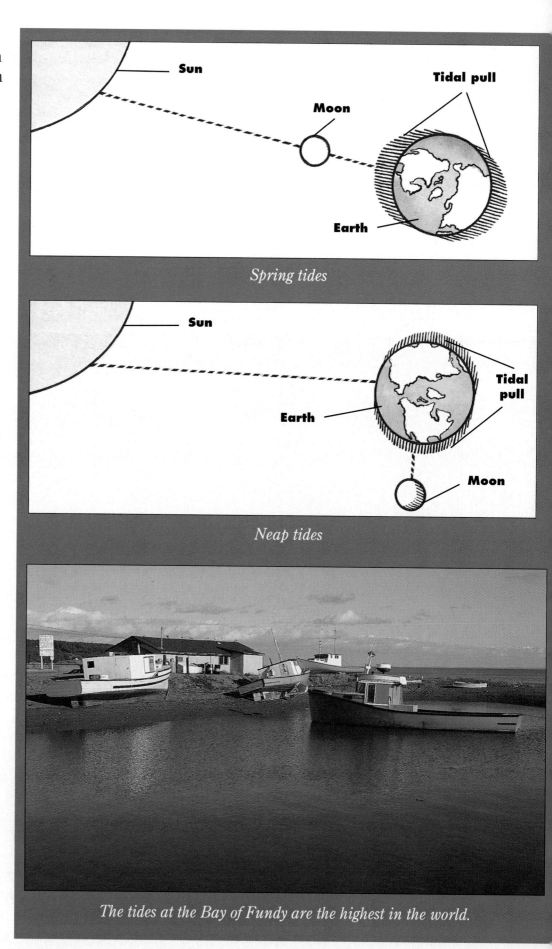

Spring tides

Neap tides

The tides at the Bay of Fundy are the highest in the world.

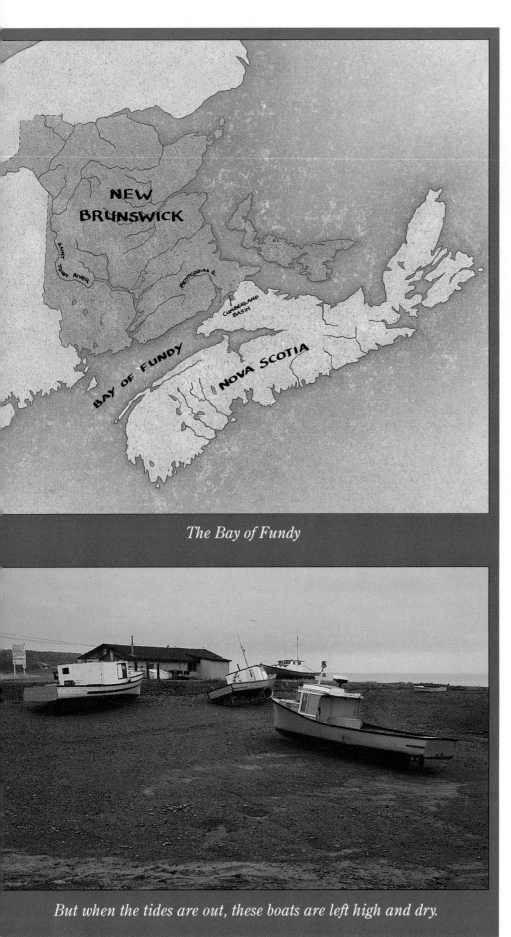

The Bay of Fundy

But when the tides are out, these boats are left high and dry.

Fundy's enormous tides cause the famous Reversing Falls at the mouth of the Saint John River. Instead of flowing downstream the way a river should, the river is forced to flow backwards — upstream — at high tide.

Tides can also rush up a river as a solid wave of water, like they do as the Petitcodiac River flows up to Moncton, New Brunswick. That solid wave is called a tidal bore. Tidal bores are caused by a shallow spot, like a sand bar, at the river mouth. It holds back the rising tide until a wave builds high enough to crash over the barrier. The bore in the Petitcodiac is small compared to the immense wall of water, sometimes eight metres high, that roars up China's Tsientang River.

Tides are more than just fascinating — they may one day be very useful to our world. As you probably know, burning oil to create electricity pollutes our environment. People are looking to the Bay of Fundy tides as a source of new — and clean — energy. After all, the tides will be here long after we've drained the last drop of oil from the Earth.

DON'T BLAME IT ON THE TIDES

A tsunami hits land.

The huge wall of water that's called a tidal wave, or tsunami (soo-NAM-ee), has nothing at all to do with the tides. Tsunamis are caused by volcanoes and earthquakes under the sea. And it's pretty easy to tell when a tsunami is going to strike land.

Scientists know that tsunamis travel at a speed of 450 to 650 km/h — about twice as fast as the fastest race car. Once scientists map out a tsunami's course, they can figure out just when and where it will hit.

One wave in 1755 measured 15 m, and it was so strong that it destroyed Lisbon, the capital city of Portugal. A little warning would have been a big help!

CHECK IT OUT!

Did you know that tsunamis occur all along the west coast of North America? At your library, find a phonebook from Honolulu, Hawaii; San Francisco, California; or Prince Rupert, British Columbia.

Does it warn about tsunamis? What does it say you should do if a tsunami is on its way? What other parts of the world do you think might have tsunamis?

BUG eyed!

WEBS *of* STEEL

How do you catch
a Boeing 747 in a net?
Make the net out of
spider silk!
That's right — the same
stuff that spiders use
to weave those sticky
webs. Spider silk is
so strong that strands
of it one centimetre thick
could make a net that
would hold a jumbo jet!

But that's only part of the
story. Not only can the web
stretch so much that it won't
even snap when hit by a speeding bum-
blebee, but it can shrink, too! There's
nothing in your home like it (except in
the corner of the ceiling where you
haven't swept).

BY JAY INGRAM

When a gust of wind blows through a spider web, it should make the web sag in the middle. Picture an elastic band stretched tight, then going limp when its two ends are pushed together. But a spider web doesn't go limp — instead, the web shortens. An elastic band can't do that. But the spider web can. You've got stretch and shrink in one amazing material.

What's the secret? Spider glue. Spiders coat the threads of their webs with tiny droplets of glue to make sure insects don't get away. Those glueballs look like miniature beads on a string. Inside the glueballs the threads are all wound up like a string around a yo-yo. If a big gust of wind blows through the web, more threads get pulled into the glueball and wound up. The web stays in its original shape.

Stretching? That's different. If a bumble-bee smashes into the web, the threads have to stretch three or four times their original length without breaking. The secret is that glue again. The water in the glue keeps the threads wet, so not only can they unravel, but the strands inside the threads can also slide past each other to stretch the web even more.

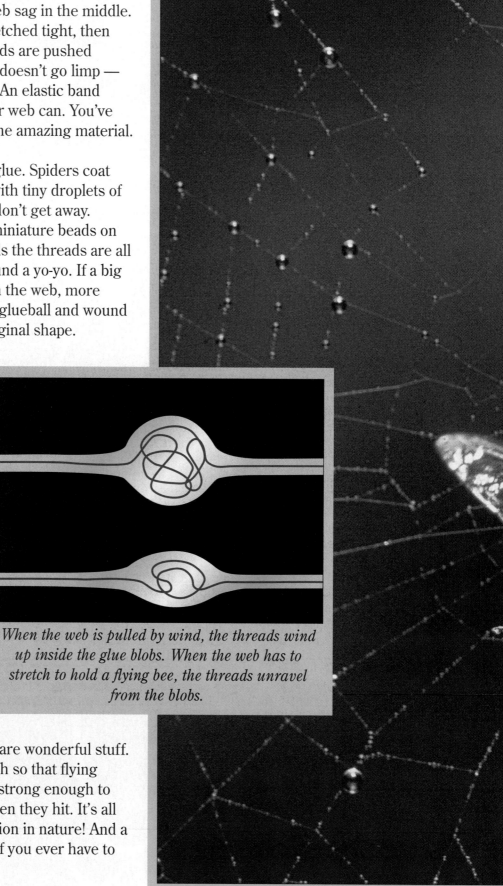

When the web is pulled by wind, the threads wind up inside the glue blobs. When the web has to stretch to hold a flying bee, the threads unravel from the blobs.

Spider glue and spider silk are wonderful stuff. The threads are thin enough so that flying insects can't see them, but strong enough to hold on to those insects when they hit. It's all biodegradable, too. Perfection in nature! And a handy tool to have around if you ever have to catch a jumbo jet!

Spiders use different kinds of thread to build each part of their webs. Use different colors of yarn to make your own spider web. Use an upside-down chair as a guide for your triangle.

Using your first color, make a triangle.

Using your second color, tie a piece of yarn from end to end of the triangle to split it in half. Then use the same yarn to split the two sections in half.

Use the same yarn to split each of the sections in half.

Use your third color to make a spiral around the center. Tie the yarn on at every place that it touches a part of the web.

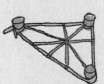

Tie a piece of your first color to the center of the web. Leave a long piece of yarn hanging. This is the tripline. Stand away from the web with the tripline in your hand. Have someone throw a piece of paper onto the web. Did you feel it? How do you think the tripline helps a spider?

Creep
Show

BY RICHARD CHEVAT

**Who hasn't seen
or heard of a
horror movie
starring a terrible, giant,
and deadly insect?
But giant ants
and killer crickets
only exist
in the movies,
don't they?**

*Berenbaum holds a tarantula
as she teaches people not to be
afraid of insects.*

If you want the facts about insects and
horror movies, there's no one better to
ask than May Berenbaum. She's a professor at the University of Illinois, and every
year she runs an "Insect Fear Film Festival,"
showing such films as *The Bees, Kingdom of
the Spiders,* and *The Giant Mantis.*

Why is a professor of entomology (the study of insects and spiders) showing horror movies? These films are meant to be scary, but by the time Professor Berenbaum gets through explaining how silly they are, audiences wind up laughing at the giant moths and ants. Before each film, she explains exactly why the monsters on the screen could never exist in real life.

"One year we showed a film called *Kingdom of the Spiders*, which is about tarantulas that take over a town, killing everybody," says Berenbaum. "Tarantulas are overrated. It's true that they're predators — animals who eat other living animals. But spiders are really shy. They're not out to get humans. As far as they are concerned, you're just a part of the landscape."

What about Mothra, the giant moth? Or the huge praying mantis that climbs the Washington Monument in the movie called *The Giant Mantis*? One of the favorite gimmicks of moviemakers is to feature insects that have somehow grown fantastically huge. But Berenbaum explains that, even if an insect somehow managed to grow that big, it could never survive, much less eat entire cities. "There's a good reason you don't see six-metre insects," she explains. "For example, if they were that big, they wouldn't get enough oxygen."

Insects don't have lungs or blood. Oxygen gets to their cells through a series of small tubes that branch from the surface throughout their bodies. If the tubes were too far from the insides of their bodies, they wouldn't get enough oxygen.

A six-metre-tall insect would have another kind of problem. "All insects molt," says Berenbaum. "They have a hard outer skeleton, like a shell. When they grow, they have to grow a new skeleton."

216

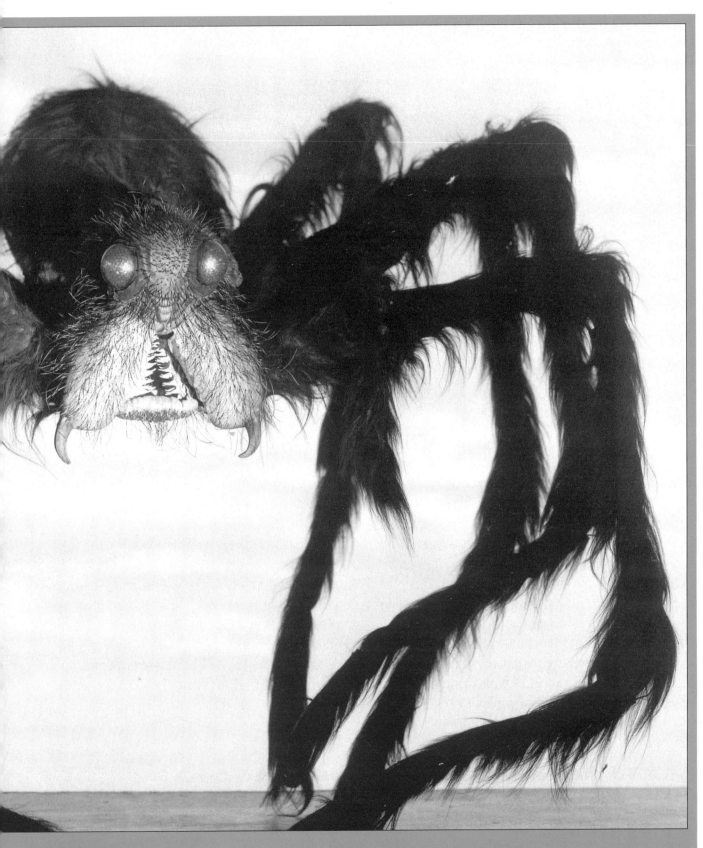

*This fake tarantula doesn't even look real — or,
for that matter, like it could eat a town!
Turn back one page to see two real tarantulas for yourself.*

217

Insects molt by shedding their old skeletons. Underneath is a new one. But right after they molt, the skeleton is soft and pulpy. It takes a while for it to grow hard. This would mean big trouble for a huge bug.

"If a giant insect molted," says Berenbaum, "it would weigh so much that gravity would flatten out its new soft skeleton as if it were made of jelly."

A 15-m ant would ruin any picnic. Fortunately, you'll never see one more than 2.5 cm long.

Filmmakers create insects that are much scarier and more dangerous than the real thing. In a film called *The Naked Jungle*, a colony of army ants attacks a plantation, eating everything in its path, including people. But in real life, Berenbaum explains, army ants eat other insects, and they only travel about 4.5 m a day. So if you see some coming toward you, there is plenty of time to step aside.

Berenbaum thinks insect horror movies are popular because many people are afraid of insects. "When I was a kid I was afraid of insects," she says. "I was always interested in biology and all kinds of animals, but not insects. When I got to college, I took a course and I realized that insects were really amazing.

"Everything about them is interesting. They're built in a way that is completely opposite from humans. Their skeletons are on the outside instead of the inside. Their nerves are wired differently. They breathe differently, they move differently, they eat different things."

You can learn all about what insects are not like by watching them in the movies. Even though you know giant ants can't exist, it's still fun to watch them on the screen.

But it's just as much fun to watch and learn about the real thing. Did you know that the smallest insect is smaller than a single human cell? Or that termites build homes that can be more than 4.5 m tall? The more you learn, the more you'll find that real insects are even more amazing than those made-up movie monsters!

CHECK IT OUT!

Find out how moviemakers make those bugs look so huge on screen. If you can borrow a video or movie camera, try using it to film bugs. Can you make them look huge? Can you make them look big enough to take over a city?

A fly's eye has over 4000 facets that let it see in any direction without moving.

TAKE a HIKE

What better way to find out about insects
in your neighborhood than by going
on an insect hike?
You can find out where lacewings live,
how caterpillars crawl, or what earwigs eat.
Grab a friend and go!

BY PAMELA M. HICKMAN

Theme hikes

How about a dragonfly hike or a grasshopper hike? If you have favorite insects, focus an outing on them. Find out what habitat they prefer and head for it. For example, dragonflies dance along the edges of ponds and streams, but grasshoppers go for long grass and wildflower meadows. When you're concentrating on only one or two kinds of insects at a time, you will be able to spend more time learning how to identify them. You can watch for features such as color, size, and wing shape, and begin to recognize different species.

Grasshopper

Some provinces and states have checklists for certain insects, such as butterflies, that list all the species of butterflies that have been found in a particular area. When you identify a butterfly, you can check the list to make sure the butterfly is really found in the area where you saw it. A list might also help you realize you've made an unusual sighting! Contact your provincial or state nature organization or natural history museum to see if there are any checklists you can use.

Monarch butterfly

Mini-hikes

Who said hikes have to be long, or even on foot? Take a mini-hike and give your feet a rest. Try crawling around on your hands and knees, taking time to really look at the ground for those well-hidden bugs. Or, let your fingers do the walking. Find a rotting log and gently probe and poke your way through it with your hands and a pair of tweezers. Take a mini-hike on a tree. Start at ground level and work your way up, checking in the grooves of bark, under loose bark, in

Earwig

holes, in buds, on and under leaves, and in blossoms, seeds, cones, or nuts. You can hike for hours and never get sore feet!

Shake it!

Small shrubs are super hiding places for insects. Wherever they're clinging, you can shake them out to get a closer look. Take an old light-colored sheet and spread it on the ground under a shrub. Give the bush a good shake. You should find that several different kinds of creatures have dropped on to your sheet.

TAKE IT ALONG

What will you need for your hike? Go over this checklist to make sure you're prepared, and make sure someone knows where you're going.

When you're through looking, leave them near the bottom of the bush and they will climb or fly back up.

Centipede

Garden slug

When you're out for a walk, play a game of peek-a-boo with the wildlife. Many insects and other animals shelter under stones, rocks, and rotting logs. Carefully lift up these hiding places to reveal the life hiding below. If you're lucky, you'll find various ants, beetles, wood lice, centipedes, millipedes, slugs, snails, earthworms, and maybe even a salamander. Don't forget to replace the stones or logs where they were so the animals will be protected.

☐ **a magnifying glass**
☐ **a notepad**
☐ **a hat to keep the sun off**
☐ **a snack if you're going to be a while**
☐ **boots if you're investigating a wet area**
☐ **long sleeves and pants if you're checking out wooded areas**

CUTTING EDGE

BITING BACK

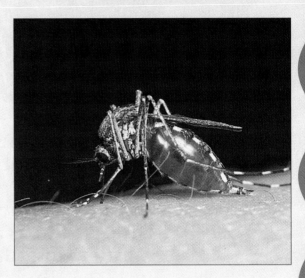

You probably don't like mosquitoes because they bite. But did you know that sometimes their bites can be a real problem? Mosquitoes can carry and pass on a disease called malaria (muh-LAIR-ee-uh). If researchers have their way, you might be able to help stop the disease.

Leon Rosenberg, a doctor at Stanford University in California, is working on a mosquito vaccine for humans. Any mosquito that bit a vaccinated human would die. And a dead mosquito can't pass malaria from one person to another!

But this is a vaccine with a twist. It wouldn't protect a person against getting malaria in the first place. So doctors are hoping that people would get the vaccine just to help other people. That might just happen. Who wouldn't want to get back at those buzzing biters?

MIGHTY
M I N I A T U R E S

BY SUSAN HUGHES

**Insects and spiders really are amazing.
They are responsible for all kinds of things,
from pollinating flowers to
eating pests that munch on food crops.
But some of them are amazing for
something altogether different —
their sheer size!**

What's the largest water insect in the world? The giant waterbug of Venezuela and Brazil. To see just how long it is, put a toilet paper roll on its side. A tropical giant waterbug can grow as long as that roll — or even longer! Hidden in the mud at the bottom of a pond, the waterbug will attack tadpoles, salamanders, and insects, jabbing at them again and again with its sharp mouth parts. This expert predator can even kill a fish twice its size.

Giant waterbug

The longest of all insects, the giant walkingstick of southeast Asia, is almost as long as two newly sharpened pencils placed end to end. A smaller version of the giant walkingstick is found in North America. It's our largest insect even though it's not half as long as its record-breaking relative! You'd have to be pretty lucky to see a giant walkingstick actually walking, though. During the day, it stays very still, camouflaged perfectly among twigs and branches. At night, the walkingstick gets busy feeding among the leaves because that's when most insect-eaters sleep.

What hangs from the branches of palm trees and can peel a banana? A monkey? Nope. It's the goliath beetle of West Africa, the heaviest insect in the world. A goliath beetle is bigger than a sparrow — and just one of its wings is wider than the palm of your hand! And those wings aren't a waste. This beetle is the largest flying insect in the world. Mind you, it does have trouble getting its great weight off the ground. It has to prepare for lift-off by beating its wings really quickly. Only after its buzzing wings have raised its body temperature and warmed up its flight muscles can this giant become airborne.

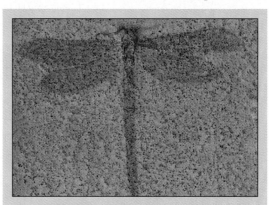

Goliath beetle

Which insect is the biggest of all time? It's definitely the prehistoric dragonfly. It probably looked like a modern dragonfly — except it was about four times as big as the biggest dragonfly you'd see today!

Take a block of butter out of your refrigerator and hold it in your hand for a minute. That's 454 g — about the same weight that scientists think this ancient dragonfly probably weighed. As it flew over ancient swamps and marshes 280 million years ago, this giant must have cast quite a shadow. Fossilized impressions of its wings suggest that, from wingtip to wingtip, the dragonfly measured about 70 cm, which is longer than your arm. Now that's making a big impression!

Above: Prehistoric dragonfly
Below: Giant walkingstick

And its air speed? Hang on to your hat. Scientists guess that this prehistoric jet could keep up with one of today's cars on the highway!

Look at the size of the butterfly on this page. It looks huge, doesn't it? Well, the Queen Alexandra's birdwing butterfly isn't quite this big, but it is big. If its wings are spread out, they reach from the top to the bottom of the lefthand page! It should come as no surprise that the Queen Alexandra is the world's largest butterfly.

Queen Alexandra's birdwing butterfly

224